Desert

The Mojave and Death Valley

Photographs by Jack Dykinga

D e

The **Mojave** and

Text by Janice Emily Bowers

s e r t

Death Valley

Harry N. Abrams, Inc., Publishers

For my dear friends Linda and Tom Caravello,
whose humor and selfless humanity are a constant inspiration.
They showed me the desert—they taught me love. — JD

Editor: Robert Morton
Designer: Robert McKee

Library of Congress Cataloging-in-Publication Data
Dykinga, Jack W.
    Desert: the Mojave and Death Valley / photographs by Jack Dykinga;
text by Janice Emily Bowers.
    p.   cm.
    Includes bibliographical references.
    ISBN 0–8109–3238–5
    1. Natural history—California—Mojave Desert. 2. Natural history—Death Valley (Calif. and
Nev.) 3. Natural history—California—Mojave Desert Pictorial works. 4. Natural history—Death
Valley (Calif. and Nev.) Pictorial works. I. Bowers, Janice Emily. II. Title.
QH105.C2D95 1999
508.794'95'0222—dc21   99–25454
Photographs copyright © 1999 Jack Dykinga
Text copyright © 1999 Janice Emily Bowers

Harry N. Abrams, Inc.
100 Fifth Avenue
New York, N.Y. 10011
www.abramsbooks.com

Title page, left: *Spine-covered red barrel cactus* (Ferocactus acanthodes) *on the slopes of Cima Dome.
East Mojave, Mojave National Preserve, California.*

Title page, right: *Cracked patterns of mud flats on valley floor, with popcorn flowers* (Cryptantha sp.)
growing along borders. Death Valley National Park, California.

Opposite page: *Saltgrass* (Distichlis spicata) *and lizard tail* (Anemopsis californica). *Warm Sulphur
Springs, Panamint Valley, Death Valley National Park, California.*

San Francisco

Sacramento ★

Salt Lake City

Fresno

Tonopah
Caliente

Lone Pine

Scotty's
Castle

St. George

Lake Powell

Stovepipe
Wells Village

Death
Valley
National
Park

Las
Vegas

Lake Mead

Hoover Dam

Grand Canyon Village

Sequoia
National
Park

M o j a v e   D e s e r t

Lake
Mojave

Colorado River

Bakersfield

Tehachapi

Mojave National
Preserve

Kingman

Flagstaff

Mojave

Edwards AFB

Barstow

Sedona

Victorville

Needles

Lake Havasu

Prescott

Lake Arrowhead

Big Bear

Joshua
Tree

San Bernardino

Los Angeles

Palm
Springs

Joshua Tree
National Park

Colorado River

Indio

Phoenix

San Diego

Tucson

CONTENTS

I

Two years ago, after living in the middle of Tucson for almost twenty years, I moved to a bit of desert on the outskirts of town. My stated reason was to get away from sirens, helicopters, and traffic and to escape the leakage of other people's lives into my own—radios, car alarms, dogs, firecrackers, lawn mowers, hedge trimmers, weed whackers. I needed silence and a little space around me. Unstated was my fear that living in towns for so long, I had become town-wise and desert-foolish. I knew where to buy real tomatoes, but not where to find spadefoot toads after the first drenching thunderstorm of summer. I could tell you how late my favorite bookstore stayed open, but not when the sun would rise the next morning or whether the moon was waxing or waning. More often than I care to admit, I was adrift in the seasons, telling myself in November that the days were surprisingly cool for May and thinking in July that April's migrating warblers should be arriving any day now.

I was tired of being a fool. I wanted to step outside at night and not be surprised by the moon. I wanted to see how much rain made desert washes run, and run hard. Most of all I wanted to insinuate myself somehow into the lives of wild animals. What form this would take, I hardly dared to plan. Henry David Thoreau's quiet communion with small animals came to mind—he reportedly summoned squirrels and songbirds with a whistle—but I was willing to be satisfied with much less.

At the end of October, I took possession of a house on an acre and a half of land. In January, my land took possession of me when I slid my pitchfork into the compost pile, lifted the dry thatch on top, and uncovered two baby cottontail rabbits in a nest. Bodies touching, they huddled rump to head, fitting together as neatly as figure and ground in an Escher drawing. Their gray ears lay back on gray-brown haunches. Neither one moved. I hardly breathed myself. This was why I had moved here. This was what I had wanted all along.

Lichen-covered volcanic rock with snakeweed (Gutierrezia *sp.*) *in lava flow near Hunter Mountain. Death Valley National Park, California.*

Shifting the center of my day by approximately twelve miles westward was like moving to another world. In my neighborhood, houses are like scattered islands in an inlet or sound, and the tide of desert rises and falls around them every day as animals come and go about their business. Watching Gambel's quail every morning and evening, I learned more about their habits in my first month at the new place than I had in thirty years of hiking through the desert. I came to recognize a couple of javelina who dropped by almost daily for a drink from the waterhole. I watched spadefoot toads call for mates as they floated across a mud puddle, and I raised some of their eggs into tiny toads the size of my thumbnail. I saw my first bobcat, and my first gray fox.

Now, more than ever, the Sonoran Desert is my home, and when I go to the Mojave Desert, I am keenly aware of being a tourist. Squeezing visits into one week here, another week there, cannot substitute for living in a place until you know it so well that its rhythms knit into your bones, until, say, the alarm call of a Gambel's quail—an anxious ticking that varies in volume, duration, and rate—conveys to you real information about what is out there (coyote? rattlesnake? cat?) and how much of a threat it poses. As a visitor, I cannot expect to acquire this level of intimacy. It is always time to move on. No time to hang around a patch of milkweeds while a monarch butterfly decides whether to dot the leaves with eggs; no time to keep an eye on a red-tailed hawk's nest until the nestlings have fledged. Of the million stories in the naked desert, I am fated not to know the end of a single one.

My acquaintance with the Mojave Desert is rather recent. As a child in California, I lived within an hour or two of the desert's edge, and I must have seen it now and then, if only on family trips to Oklahoma where my aunt and uncle lived on a farm. I remember that on one such trip we ate a picnic lunch at a public park in Needles, California. There was nothing about the town to mark it in my mind as desert except perhaps for the wind. Strong gusts had scoured the streets of trash and dust, and as we drove along the main avenue, old-fashioned Christmas decorations—swags of red and green tinsel twisted into curlicues and wreaths—swayed overhead, and red plastic bells trembled soundlessly on lamp posts.

I also remember a trip to the ghost town of Calico, California with my Girl Scout troop. I wore my

Kodak camera on a long, black strap around my neck and snapped pictures of my friends clowning on the porches of rickety wooden buildings. That once upon a time Calico had been a real place, not a faked-up Wild West for tourists, did not matter to me then, and that it existed within a larger framework called the Mojave Desert made no impression on my consciousness. Even then, in the 1950s, California's population was growing rapidly, but my memories do not reflect the boom. Towns were small and decently separated from one another by fields or vacant lots. The best strawberries in the world were grown by Nisei farmers in a narrow plot of land between our town and the next. My world did not seem crowded except at the Los Angeles County Fair, where you expected to see people and lots of them, and at the beach, which is one reason my family went to the mountains for summer vacations. We could have gone to the desert but for some reason did not; evidently the desert was a place you drove through, not to.

Things are different today: more people, more houses, more cars. As more land disappears under asphalt and concrete, the remainder suffers greater pressure for recreation and diversion. Four million people now live within a day's drive of the Mojave Desert, and when *Sunset* magazine or the *Los Angeles Times* tells them that wildflowers are spectacular in Death Valley this spring, half of them get in their cars to check it out, less because they love wildflowers or deserts than for something to do.

Sometimes I don't know which group worries me most: the millions who visit or the millions who stay away.

Among the millions who stay away there must be some who would not benefit from a visit, who indeed should be encouraged not to come. In this group I would put developers, cactus collectors, off-road-vehicle users, and anyone else who treats the desert as a consumer commodity. Others stay away from simple lack of interest, which is fine: the world would be a dull place if we all liked the same thing, as some wise soul has observed. When Lack of Interest casts (or fails to cast) its ballot on election day, however, the results can be devastating to those who really care.

The millions who visit the Mojave Desert cannot be accused of not caring. But, choosing to care in comfort, they have brought about a vast and expensive infrastructure to meet their needs for food, water, toilets, beds, showers, gasoline, film, newspapers, cigarettes, ice, sunscreen, decongestants, and videos, not to mention the ubiquitous souvenirs that could, given minimal alteration, just as well be sold on Cape Cod: decorative teaspoons, ash trays, coffee mugs, snowglobes, stuffed animals, and so forth. Some visitors bring enough water, food, and bedding to pretend that they are independent of infrastructure, at least for a few days; most of the rest are unprepared for even the simplest emergency, and if a radiator hose breaks, they would die if not for the cell phone that summons a wrecker from two hundred miles away.

Whether prepared or not, visitors need paved roads and parking lots, motels and quick-marts, which is to say that they need vegetation to be bladed, dirt to be hauled, gravel to be laid, asphalt to be poured. They need trenches for pipelines, both water and natural gas, and borrow pits so that highways and railroads can be built on elevated beds. They need shade trees around parking lots to blunt the heat that rises from black asphalt.

All these needs and many others are met and paid for. The paying is essential, of course. Visitors pay out of pocket and through taxes for the infrastructure that supports their travels into the Mojave Desert and out again. It was in fact their willingness to pay that brought much of the infrastructure into existence.

But there are hidden costs for which no one pays. Animals are displaced when bulldozers arrive. Shade trees attract alien birds such as starlings and house sparrows, which aside from being noisy and messy, tear apart the nests of native birds to get nest sites for themselves. Weed seeds travel along roads, filter down washes, spread into untouched desert, crowding out native grasses and flowers and increasing the risk of wildfire. Pipelines are also corridors for weeds. Roads and railroads, in crossing vast plains between mountains, throw up a barrier to flowing water, depriving plants on the downstream side. Trash bins and landfills, a smorgasbord for ravens, support a many-fold increase in raven numbers, growth that works to the detriment of baby desert tortoises whose soft scutes make them vulnerable to those big, hard black beaks.

Over and above the generalized damage to which every visitor in some way contributes, there is the particular damage that individuals do from ignorance or malice. At the tackier gift shops you sometimes see tarantulas and scorpions embedded in clear plastic. There's malice for you. Those animals had to come from somewhere, and they did not, believe me, come from an arthropod ranch: they were trapped in numbers at bait stations. Contrary to popular myth, our desert tarantulas are not dangerous to humans, nor do scorpion stings threaten the lives of healthy adults. These are interesting and beneficial animals. If left unmolested, a female tarantula can live to be twenty-five years old. Desert scorpions quietly consume cockroaches and crickets, subduing them first with a venomous jolt. (No one, not even a scorpion, likes a meal that wiggles.)

"Wait a minute," you say, "I've never killed a tarantula in my life. I even swerve to avoid them on the highway." Well, try this on for size. Perhaps you have caught and kept a horned lizard or dug up a flowering cactus plant, telling yourself, "There are plenty more out here." Sorry to say, I have too, using that rationalization and others. The problem is that there are always plenty more until suddenly there are not. Ignorance is more pervasive than malice and far more damaging in the long run.

The first step in repairing ignorance is to stitch it up with a few strong facts. Here are some facts about the Mojave Desert. The smallest of the four deserts in North America, the Mojave Desert is about 35,000 square miles, roughly the size of Indiana or Maine. It is a place of rugged mountains, long valleys, and tip-tiited alluvial fans. Elevations range from 282 feet below sea level to about 4,000 feet above sea level. The Mojave Desert is home to a number of threatened animals, among them the desert tortoise and the desert horned lizard, and to about one hundred rare species of plants. Small sections of the desert—active dunes, for instance, and saline flats—are barren of plant life. Otherwise, the Mojave possesses a varied and wonderful flora, from groves of wild palm trees that depend on springs for sustenance to vast monocultures of creosote bush that can go for a year or longer without rain.

And sometimes they must. At low elevations, rain averages an inch or two per year. As elevation increases, so does rainfall; even so, few parts of the Mojave Desert receive more than eleven or twelve inches annually. At about 4,000 feet above sea level, precipitation is high enough that desert scrub gives way to an open woodland of piñon and juniper. As you might expect, the Mojave Desert can be hot in summer: temperatures of 120° degrees F. are not uncommon. (The record, 134° F., should be hot enough for anyone.) Freezing nights are frequent in winter, especially in valley bottoms where cold air collects, and at higher elevations.

Yet facts alone do not constitute knowledge; we also need interpretation. And, when it comes to deserts, which many people seem to hate on principle, we need sympathetic and informed interpretation. In 1894 an imaginative journalist informed readers of the New York *World* that Death Valley was "a pit of horrors—the haunt of all that is grim and ghoulish. Such animal life as infests this pest-hole is of ghastly shape, rancorous nature and diabolically ugly." To me, these sound like the words of someone who never went near Death Valley. I suspect that the writer was inventing from a small base of scientific illustrations and travelers' tales. I also suspect that he was afraid of the desert: afraid of what it might contain, afraid of what could happen, afraid that he himself could not survive there.

The French have a saying: *tout comprendre tout pardonner;* to understand everything is to forgive everything. In the desert, where much is strange to human eyes, a little understanding can ease one's natural fears. Many people are taken aback by their first sight of a chuckwalla, for example, and from an aesthetic point of view, there's no doubt that this large, ungainly lizard leaves something to be desired. Not only is it broad in the beam and rather flabby looking but it sheds its skin in patches, as if recovering from a bad sunburn. Unattractive, I'll grant you, but diabolically ugly? I think not. Nor is it the least bit ghoulish; in fact, the chuckwalla shuns meat and subsists solely on leaves, seeds, flowers, and fruits. Far from having a rancorous nature, the chuckwalla is a timid creature and when threatened by a predator is likely to back into a crevice and inflate its body to wedge itself in place.

What the chuckwalla needs—what the entire Mojave Desert needs—is a better press agent. That is in part the purpose of this book—to make new friends for the Mojave Desert, and to remind old friends that despite the changes they deplore, the desert still needs their good will. For that reason, you will find no harangues or tirades here, no bemoaning or bewailing. Our job, mine and Jack Dykinga's, is not to break your heart; our job is to show you why we love the Mojave Desert, and why you might love it, too.

Creosote bush (Larrea tridentata) *clings to rim of gorge in afternoon light. Piute Gorge, Piute Mountains, Mojave Natural Preserve, California.*

Cottontop cactus (Echinocactus polycephalus) *glows in sunset light. Cinder Cone Lava Beds, Mojave Natural Preserve, California.*

The view from Toroweap Overlook of the Grand Canyon, as morning light filters through the place where the Mojave Desert in the canyon meets the Colorado Plateau on the rim. Grand Canyon National Park, Arizona.

Tufa formations rising from ancient Searles dry lake, at dawn. Trona Pinnacles Recreation Area, Bureau of Land Management, California.

Eroded granite boulders of the New York Mountains at dawn, with crescent moon rising. Mojave National Preserve, California.

Saline Valley's southern end, with weathered granite boulders near Hunter Mountain, and the Nelson Range in the background.
Death Valley National Park, California.

Crumbling granite boulders of the Eastern Mojave Desert's Coxcomb Mountains, at dawn. Joshua Tree National Park, California.

Mound cactus (Echinocereus triglochidiatus *var.* mojavensis) *at sunset, with Joshua tree* (Yucca brevifolia). *Near Kessler Peak, Mojave National Preserve, California.*

Alluvial boulders on valley floor, with lone desert sunflower (Geraea canescens) *beginning to bloom. Death Valley National Park, California.*

Mojave prickly pear (Opuntia phaeacantha) *with a dusting of snow and piñon pine cones* (Pinus monophylla). *New York Mountains, Mojave National Preserve, California.*

Opposite: Dying blades folded down the stalk of a Mojave yucca (Yucca schidigera). *Mojave Natural Preserve, Piute Creek, California.*

The Calico Hills, with sunset light framed by silhouetted Mojave yucca. Red Rock Canyon National Conservation area, Nevada.

Barrel cactus, Mojave yucca, and cholla, at dawn. Mojave National Preserve, Piute Range, California.

Teddy Bear Cholla (opuntia bigelovii) *from the Sonoran Desert and Mojave yuccas* (Yucca schidigera) *from the Mojave Desert, Sacramento Mountains, California.*

Panamint Valley with hillsides covered with desert trumpets (Eriogonum inflatum) *punctuated by creosote bush* (Larrea tridentata). *Panamint Mountains, Death Valley National Park, California.*

Scattered volcanic debris in the sandy flats surrounding Eureka Sand Dunes. Death Valley National Park, California.

Salt formations on valley floor, with dawn's light on the Panamint Range. Death Valley National Park, California.

Dawn on the Panamint Mountains, with salt evaporation patterns in the valley floor. Death Valley National Park, California.

Crystallized salt formations with Amargosa Mountains in the background. Death Valley National Park, California.

Cracked clay amid the Mesquite Flat sand dunes with creosote bush (Larrea tridentata) *and the Grapevine Mountains in background. Death Valley National Park, California.*

Cracked valley floor, with intricate salt patterns and the Panamint Mountains in the background. Death Valley National Park, California.

Gray Riccardo formation sandwiched between red volcanic cap rock formations in Hagen Canyon Natural Area. Red Rock Canyon State Park, California.

Opposite: Weathered, eroded sandstone, with iron oxide and calcium carbonate coloration known as Aztec sandstone. Whitney Pockets, Nevada.

Ubehebe Crater cinder flats with islands of desert holly (Atriplex hymenelytra) and the Dry Mountains in the background. Death Valley National Park, California.

Opposite: Lava flow amid cinder cones with globemallow (Sphaeralcea ambigua) growing amid jagged lava formations. Mojave National Preserve, California.

II

John C. Van Dyke, an art historian and aficionado of arid places, wrote in 1901 that the desert was "no place for flowers." Shrubs and trees bloom on occasion, he noted, and "many tales are told of the flowers that grow on the waste after the rains, but I have not seen them though I have seen the rains." He dolefully cataloged what he had missed: "no lupins, phacelias, pentstemons, poppies, or yellow violets" and added that the few species he had noticed "do not make up a very strong contingent."

There is no reason to disbelieve him. He likely endured drenching summer thunderstorms—the rains of which he spoke—but the "flowers that grow on the waste," the lupines and phacelias he did not see, are spring-blooming annuals that germinate only after plentiful autumn or winter storms. Rain of an inch will do the trick if it falls between late September and early December. Unfortunately for Van Dyke, his travels across the Sonoran and Mojave deserts from 1898 to 1901 coincided with a series of very dry winters. Not only were there no one-inch rains but the combined storms of September, October, November, and December did not even add up to an inch in 1898 or 1899. If he had been a year or two earlier, he might have had better luck: but he was not, and for him the desert remained no place for flowers.

Having endured summer downpours, Van Dyke knew that rain could work swift changes in the desert. "In the canyons the swollen streams roll down boulders that weigh tons," he wrote. "In a very short time

California Poppy Reserve: California poppies (Eschscholzia californica), *rabbitbrush, red owl's clover* (Orthocarpus purpurascens), *goldfields* (Lasthenia chrysostoma). *Antelope Valley, California.*

there is a great torrent pouring down the valley—a torrent composed of water, sand, and gravel in about equal parts." A two-hour deluge can not only reshape a river channel, it can resurrect an entire flora and fauna from its summer torpor. Cinchweed, thornapple, devil's claw, spiderling, and summer poppy burst into bloom. Dragonflies, tarantulas, tortoises, ground squirrels, spadefoot toads, mites, millipedes, lizards, beetles, and butterflies crawl, fly, dig, hatch, or stumble out of hiding, spend a few days or weeks renewing acquaintance with the desert and with one another, then disappear until the next time. Don't blink, or you'll miss it. Don't assume that it will be the same next week or next year, because it won't.

The same applies to springtime in the desert, as Van Dyke discovered for himself. That desert springs vary from year to year is not a closely guarded secret, yet for some reason the word has failed to spread. As a botanist working for the federal government, I get calls every year from journalists, photographers, editors, and others wanting to know when spring wildflowers will reach their peak. The assumption seems to be that wildflower displays in the desert are as predictable as ice in Antarctica and that seeing them is merely a matter of reserving airplane tickets for mid-March as opposed to late April.

Reactions range from dismay to incredulity when I deliver the bad news that there has been no rain for three months and that this spring wildflowers will live only in *Arizona Highways*. One photographer kept me on the phone for a good ten minutes after I said "Nope, not this year." He could hardly believe that a mere civil servant (although becoming less civil by the minute) could foil his plans. If pressed hard enough, I would surely confess that wildflowers bloomed somewhere in the desert. They had to: he had already cleared his calendar and reserved his rental car.

The oddest thing about good springs in the desert is actually not how rare they are but how little we know about them. The general outline is well understood, of course. Substantial autumn and winter storms—

that crucial inch between late September and early December—can trigger germination of several hundred wildflower species; exactly which species depends in part on the temperature when it rains, some species preferring warmer or colder conditions than others. Seedlings quickly grow into rosettes—leaf clusters pressed more or less flat against the ground—then do nothing much for the next several months. With warming temperatures in spring, the rosettes bolt like lettuce plants in a vegetable garden, producing flowering stalks, flowers, and eventually seeds. After the seeds ripen and fall to the ground, they can remain dormant for many years. (If their seeds could not survive the multi-year gap between germinating rains, desert wildflowers would become extinct.)

These are the generalities. What we lack are specifics. For instance, seeds of chia and desert sunflower require scorching heat in the summer if they are to germinate the following winter. But of how many species this is the case, we cannot say. Because the germination requirements of only a few desert wildflowers have been studied in the laboratory, we do not know precisely what combinations of rain and temperature bring about gorgeous mixtures of wildflowers such as the magenta, orange, and blue of owl's clover, California poppy, and lupine.

To wildly paraphrase Tolstoy, disappointing springs are all alike (except that some are more disappointing than others), whereas good springs differ in many ways. The number of variables that might come into play are nearly infinite. Sometimes, for instance, wildflowers will germinate if a light rainstorm of a quarter of an inch follows even lighter rains that have moistened the soil crust. Another factor in some places is that non-native weeds such as Tournefort mustard, red brome, and Mediterranean split grass elbow out spring annuals by preempting space, water, and nutrients. Also important are how many and what kinds of seeds have been eaten by rodents and ants since the last good spring and how many years ago that good spring happened.

The most inscrutable factor of all is where we are in the cycling of El Niño years. The El Niño phenomenon involves differences in sea-surface temperature and sea-level pressure on opposite sides of the globe and how those differences affect atmospheric circulation. When the difference in temperature or pressure between Darwin, Australia, and Tahiti is strongly negative, we have an El Niño year. For Southern California and the Southwest, this often means a wetter-than-normal winter and spring and, perhaps, spectacular wildflower displays. When the pressure or temperature difference is strongly positive, we have a La Niña year, which typically gives us drier, colder winters and little in the way of wildflowers. The correspondence between El Niño years and good springs is not one-to-one, partly because some El Niño years are not particularly wet and partly because many other variables affect the quality and quantity of annual wildflowers. Nevertheless, I believe that the best springs are more apt than not to coincide with El Niño years.

Like seasons and years, El Niño is cyclic in nature. Short cycles on the order of four or five years spin inside longer cycles that last a decade or more. During the 1970s, 1980s, and 1990s, El Niño visited frequently, bringing so many wet winters that long-time desert residents contemplated purchasing canoes for getting to and from work. During those decades, I saw eleven good springs, approximately one every three to four years. This rate seems normal and unremarkable to me. But my perspective is undoubtedly skewed by the frequent recurrence of El Niño years during this time. In the 1940s, 1950s, and 1960s, El Niño mostly stayed away, and elderly botanists who were active plant collectors then sometimes remark that the recent spate of bountiful springs has spoiled the younger generation. They know with certainty, these crabbed and aged mentors, that good springs come only once or twice a decade, if that; moreover, they remember with painful clarity the drought of the 1950s, when for three consecutive years there was hardly enough rain to justify possession of an umbrella, much less a canoe.

I don't mind being spoiled, especially in springs like the last, which was so fantastic that it made the *New York Times.* More accurately, El Niño made the *Times,* and spring rode in on El Niño's sopping shirt-tail. A friend sent me a clipping of the article, which quoted a tourist from Oregon as saying that a spring like this happens only two or three times a century. The fact was wrong, but the feeling was right: a good spring, like the kind of emotional cataclysm that strikes at deeply ingrained patterns of behavior and belief and that happens only once or twice in a lifetime, changes everything.

In Death Valley, for instance, the main highway ordinarily traverses one of the driest, stoniest, stingiest deserts in North America, but this past spring, rockscape became bloomscape under the transformative touch of a little rain and a lot of theretofore dormant seed. It was the kind of spring that draws tourists from thousands of miles away. Wherever possible, cars were parked beside the roadway in twos and threes, having disgorged their passengers into a carnival of color. People waded knee-deep through mixtures of pink, magenta, blue, lavender, violet, white, ivory, and every imaginable shade of yellow. Some paged through wildflower guides, perhaps for the first time in their lives, and I almost envied them their ignorance: there is only one first time for everything, and the first time of seeing the desert in bloom is almost painful in its intensity, as if every flower were a needle piercing your skin or a chunk of ice pressed onto your flesh.

I wanted to join that spring, as if it were a circus, and travel with it forever. How else to experience a phenomenon as colorful as a carousel and far more restless? While native bees performed acrobatics in the blossoms, ants created intricate dances on the gravel underneath, and pairs of butterflies spiraled upward, mirroring one another's movements like vaudeville comics. One tiny black bee had filled her pollen baskets with blue pollen from blue anthers. The ants, I imagined, were asking one another, "What the heck happened to all the seeds?" Snake-head, a beautiful dandelion indigenous to the desert, hesitantly explored the limits of season and place. Each plant sent up two and only two flower heads at a time; when

these two prospered, it hazarded two more, and so on, until eight or ten heads had blossomed and set seed, rewarding caution well beyond its usual merits. Sand verbena sprawled over drifts of sand at roadside. From the flowers wafted a sweet and spicy fragrance that put me in mind of satin pillows, silken sheets, richly embroidered tapestries, and floating candles in tiled pools—perhaps the kind of setting where Sheherazade spun tales for King Shahriyar. Unlike the king, I have little need for a live-in storyteller, but if I had his wealth and power, I could imagine hiring some handsome young man to stand outside my bedroom one night and fan the scent of sand verbena through open windows.

The fragrance appealed not only to me. At dusk, long-tongued moths hovered a few inches above the lavender flowers; their tongues seemed to be hair-fine wires holding the small bodies and blurred wings in midair. By manufacturing a delectable, somewhat musky odor, sand verbena attracts insects (mostly moths and butterflies) that have learned to associate the fragrance with nectar. With every plunge into a flower tube, their tongues transfer pollen from the previous flower they visited, and if one or two pollen grains are pressed onto the pistil, eventually to produce viable seeds, the insect earns its nectar reward.

Driving through Death Valley last spring, it was hard, if not impossible, to remember that sexual reproduction was the entire point of floral display. For mile after mile, yellow flowers embraced the highway, caressing every dip and turn. The yellow bands, composed almost entirely of desert sunflower, must have been at least a mile wide, and they stretched from the salt pan in the center of the valley to the bedrock hills at valley's edge.

While not exactly a weed, desert sunflower has certain weedy traits such as wide distribution and tolerance of poor soils; moreover, the yellow color of the heads is the most common of flower colors. This is a plant that practically begs to be taken for granted. But for once I could not ignore it. I would have had to be blind. Driving through the gilded corridors of Death Valley soon had me hallucinating that the yellow center

line on the highway was not paint but an unusual alignment of flowers. I caught myself dreaming that the landscape was a giant buttercup, and everyone and everything within it had taken on a buttery glow.

This past spring was especially sweet because drought had pinched the Mojave Desert for several years in a row. You could see its results no matter where you went: collapsed or disused burrows of kit fox, kangaroo rat, and ground squirrel; hardly any pellets of cottontail or jack rabbit, all the signs of depleted animal populations. Creosote bush, the most abundant and drought-tolerant shrub of the Mojave Desert, had suffered too; most plants had lost all their leaves and about one-fourth of their stems to drought. Many young creosote bush plants had died. Now that the drought had broken, the survivors flowered with an air of having pulled a rabbit out of a hat.

At the end of one flower-filled day, a splash of yellow against a black boulder caught my eye. Most unexpectedly, the yellow turned out to be not sundrops or desert dandelion or some other wildflower but a patch of crustose lichen. Mine was a doubled surprise, in a way, because at any other time of year—or in any year but this one—lichen was exactly what I would have expected to see. There's a poem in this, I thought, a haiku about that startling reversal of expectation and color. Expectations turned on their heads, or doing handsprings and cartwheels. Yellow lichen instead of yellow flowers: the reversal seemed to be a metaphor for this wonderful spring in which the parsimonious desert had become as generous with blossom as a garden center or a flower farm.

Brown-eyed evening primrose (Camissonia claviformis) *and Arizona lupine* (Lupinus arizonicus) *with the Calumet Mountains in the background. East Mojave Desert, Bureau of Land Management, California.*

Arizona lupine (Lupinus arizonicus) *amid smoke* trees (Psorothamnus spinosus) *in the arroyos of the Sheep Hole Mountains. East Mojave Desert, Bureau of Land Management, California.*

Desert sunflower (Geraea canescens) *growing in the cracked alluvial clay soil of the Amargosa Mountain Range. Death Valley National Park, California.*

Desert sunflower (Geraea canescens) *growing in the overflow of the rain-swollen Amargosa River. Dumont Hills, Bureau of Land Management, California.*

Tufa formations rising from the shore of ancient Searles Lake, with sunset light on desert scrub. Trona Pinnacles Recreation Area, Bureau of Land Management, California.

Birdcage evening primrose (Oenothera deltoides) *in unnamed sand dunes at sunset, with Ship Mountains in background. East Mojave Desert, Bureau of Land Management, California.*

Blooming birdcage evening primrose (Oenothera deltoides) *and skeleton of remnant dried plant. Mojave Desert, Bureau of Land Management, California.*

Birdcage evening primrose (Oenothera deltoides) in unnamed sand dunes. East Mojave Desert, Bureau of Land Management, California.

Flowering prickly pear (Opuntia phaeacantha). *Red Rock Canyon National Conservation Area, Nevada.*

Grizzly bear prickly pear (Opuntia erinacea, *ssp.* erinacea) *with winding Mariposa lilies* (Calochortus flexuous) *in flower. Beaver Dam Wilderness, Arizona.*

Mojave yucca (Yucca schidigera) *in bloom, with the Granite Mountains in the background, at dawn. Mojave National Preserve, California.*

Opposite: Mound cactus (Echinocereus triglochidiatus *var.* mojavensis) *amid Mojave yuccas* (Yucca schidigera). *Cima Dome, Mojave National Preserve, California.*

Mound cactus (Echinocereus triglochidiatus *var.* mojavensis) *amid blue yuccas* (Yucca baccata). *Cima Dome, Mojave National Preserve, California.*

Coyote gourd (Cucurbita palmata) *with year-old gourds mixed among the new vine leaves. Mojave National Preserve, California.*

Arizona lupine (Lupinus arizonicus) *amid smoke trees* (Psorothamnus spinosus) *in the arroyos of the Sheep Hole Mountains. East Mojave Desert Bureau of Land Management, California.*

Opposite: Brittlebush (Encelia farinosa) *growing from granite gorge of Diamond Creek. Grand Canyon West-Hualapai Indian Reservation, Arizona.*

Flowering brittlebush (Encelia farinosa) against red sandstone formations in the northeast corner of the Mojave Desert. Lake Mead National Recreation Area, Nevada.

Flowering cinch weed (Pectis papposa) *amid blue-black cinders in the Cinder Cone Lava Beds. Mojave National Preserve, California.*

California poppies (Eschscholzia californica) *with filaree* (Erodium cicutarium) *covering the hillsides. California Poppy Reserve, Antelope Valley, California.*

Red owl's clover (Orthocarpus purpurascens) *amid carpet of goldfields* (Lasthenia chrysostoma). *California Poppy Reserve, Antelope Valley, California.*

Creosote bush (Larrea tridentata) *with flowering desert cassia* (Cassia armata) *amid branches near Cima Dome. Mojave National Preserve, California.*

Opposite: Notch-leaf phacelia (Phacelia crenulata) *amid sun-bleached desert holly* (Atriplex hymenelytra) *in alluvial debris of the Amargosa Mountain Range. Death Valley National Park, California.*

Many years ago, I flew from Algiers to Niamey, Niger, in a Boeing 747. Our flight path took us south over the Sahara Desert. I had seen the Sahara delineated on maps and globes, of course, but had never comprehended how big it is. For hours and hours, we flew over dunes. Sand extended to the horizon in every direction, a vast and apparently undifferentiated tract except when the plane's own shadow, undulating over crests and swales, differentiated it for us. Otherwise, we might as well have been flying over the ocean: no farms, roads, or towns; no woodlands, rivers, or parks. Just sand and sunlight and then, as dusk gathered, just sand.

The Mojave Desert is different. Sand is a scarce commodity. You can make any number of commercial flights and never see a dune field. In fact, you can cross the Mojave by road from north to south and from east to west and, depending on your route, never drive past dunes. If you scraped up all the dunes in the Sahara Desert and piled the sand in southeastern California, you could cover the Mojave to a depth of 1,800 feet and still have enough sand left to blanket the state of Texas. Alternatively, you could take all the dune sand in the Mojave Desert, sprinkle it evenly across the Sahara, and you would not be able to perceive that a single particle had been added. Might as well look for a grain of salt in a mixing bowl filled with sugar.

This is not to say, however, that the Mojave Desert has no dunes. By convention, topographic maps show dune fields as stippled blotches. Cartographers of the Mojave Desert apparently suffer from repetitive stress injury: how else to account for their miserly use of stipples? I want to grab a pen and redesign the entire desert. Dunes at Silver Lake north of Baker. Dunes beside the malpais at Cima Dome—black lava, white sand, a winning combination all over the globe. Dunes along the interstate, a chain of barchans just deep enough to bury the billboards. Frozen dunes on Telescope Peak, speedy dunes in the vicinity of The Racetrack. Design completed, I would turn over the project to the engineers, whose most difficult task would not be putting the sand in place but keeping it there.

Mesquite Flat sand dunes at sunset, with the Grapevine Mountains in background. Death Valley National Park, California.

The nature of sand is to move. For loose sand to form dunes, therefore, three conditions are necessary: a source of sand, a source of wind, and a place for sand to accumulate. In the Mojave Desert, the immediate source of sand is—or was—valley-bottom lakes. During the last ice age several million years ago, the landscape must have looked something like the Pacific Northwest—wooded slopes, tumbling creeks, and stairstep lakes in elevational sequence—Mono Lake draining into Owens Lake, Owens into Lake Searles, Searles into Lake Panamint, Panamint into Lake Manly, Manly into Lake Mojave. On mountain slopes, granites and other quartz-rich rocks eroded into grains of sand that tumbled with the creeks into the lakes, then settled out onto the valley floor.

Drying of the lakes about 8,000 to 12,000 years ago exposed the sand to wind, the second necessary condition. It seems likely that then, as now, wind was not in short supply in the Mojave Desert. Merely take the lens cap off your camera if you want a breeze; twist the focus ring if you want a full-blown gale. For making dunes, the trick is to allocate the abundance. After the lakes dried up, prevailing winds swept recently exposed sand from valley bottoms, piled it nearby. Countervailing winds kept it from dispersing across a broad area. Where countervailing winds were weak or nonexistent, the sand kept traveling until it reached some kind of barrier, usually low hills or mountains, then gradually trickled upslope. You see these climbing dunes frequently in the Mojave Desert, drifts of sand curled around outcrops and sprinkled over talus; where they are climbing to and whether they will arrive is something no one knows.

There is nothing like a giant sand box to bring out the child in a person. I typically arrive at a dune field for some serious purpose: to make plant lists, to assess seedling survival, to measure water content of sand at different depths. My day pack is stuffed with notebooks, pencils, field guides, and plastic bags and containers of various sizes. But at some point I inevitably find myself poised on the crest of a dune and am unable to resist temptation. I drop my pack, remove my shoes and socks, then fling myself downhill and run as fast as I can. The necessity to extricate one foot from ankle-deep sand before plunging the other foot forward hampers me, but even if I trip and fall, I will be fine, and the dune will be fine, too. No scrapes or sprains or broken bones, just sand in my hair; no environmental damage except for some dents that will disappear in the next windstorm.

At certain places on the Kelso and Eureka dunes, the sand sings with every step. The combination of compression and slippage evidently sets up an audible vibration that sometimes sounds like a muffled sonic boom, other times like tennis shoes squeaking on a basketball court. Wind shear can elicit similar sounds. So-called singing, booming, or squeaking sands occur throughout the world, but not all dunes are naturally musical. The critical requirement seems to be a surface layer of silica gel on every grain of sand. The silica might help sand grains stick together, and this cohesive mass of sand perhaps acts as a resonator when agitated by feet or wind. Whatever the physics of the sound, it seems loud when it emanates from beneath my feet—much louder than boots pounding across a wooden bridge, for instance—but for some reason does not travel far. When I reach the flats, I ask my companions, "Did you hear that?" and no one ever has.

When these amusements pall, I play at finding stories imprinted on the sand. I'm no tracker, but over the years I've picked up a thing or two from books and wildlife biologists, and under the right circumstances, when the tracks are fresh and clearly registered, I can tell bobcat from mountain lion, coyote from bobcat, and any of those from gray and kit fox. That's about the extent of my tracking skills. For example, on a dune field somewhere in the Mojave Desert I found three sets of rabbit tracks side by side. All were headed the same direction, and all ended abruptly, leaving me to imagine one rabbit making three sorties on foot and two return trips by air, or three rabbits running side by side until they shot into hyperspace. As a tracker, I make a pretty good botanist.

Last October at the Cadiz Dunes, I picked up what looked like kit fox tracks and followed them for a mile or so across the sand. The tracks headed for an expanse of hummocky, wind-rippled dunes. While foraging for seeds the night before, kangaroo rats had left abundant prints among the ripples. Vertically aligned dashes showed where tails had touched the sand, and paired dots astride the dashes marked the placement of the feet. Although capable of prodigious leaps, kangaroo rats often run prosaically on all fours. Standing near camp one night, I trained a flashlight on a kangaroo rat as it scurried about, hunting for any crumbs I might have dropped. Only once did it hop: having somehow remained oblivious of me, it ran right up to my feet, did a kind of double take—yikes!—and leaped upward and backward about twelve inches. Kangaroo rats are a favorite food of kit foxes, and I was not surprised that my fox had visited each of three kangaroo rat burrows in turn. "Hello? Anybody home?"

After leaving the hummocks, the fox had put on a burst of speed for no reason that I could ascertain, then slowed again to a walk. Another pair of tracks, also kit fox but not as well defined, joined the first and paralleled them for many yards. Two foxes on the same day or the same fox on two different days? Puzzling over the evidence, I could have been Winnie the Pooh informing Piglet that "it is either Two Woozles and one, as it might be, Wizzle, or Two, as it might be, Wizzles and one, if so it is, Woozle." Fortunately for my sense of dignity, there was no Christopher Robin to clamber down from his perch in a big oak and tell me that I was a silly old Bear.

The parallel sets of tracks continued onward, crossing a low-lying area of crusty sand where one of the foxes had plunged through the roof of an old rodent burrow—deep hole, guffaws of laughter from the other fox. Naturally, I promptly dropped into the basement of the same burrow system, sinking my right leg to the shin in sand and making an even deeper hole. There was no one around to laugh at me, so I laughed at myself, retrieved my leg, and trotted after the foxes until I lost their trail.

Later that same day I found the unmistakable trail of a sidewinder rattlesnake going up one hummock and down the other side. Sidewinders often (but not always) travel sideways to make forward progress, and the tracks they leave are distinctive parallel lines, each slightly hooked at one end. To crawl, the sidewinder steers its upper body sideways, carving a shallow channel into the sand. The lower body follows, flowing in the channel; meanwhile, the upper body lifts into the air and reaches like a tendril in the opposite direction. The head lowers to the sand and, followed by the upper body, smoothes out another channel parallel to the first. This delicate waltz enables sidewinders to get good traction on loose sand; otherwise, I imagine, the poor things would be slipping and sliding as if on ball bearings.

The sidewinder trail on the Cadiz Dunes was recent enough that I could see the imprint of individual ventral scales in the sand. Eager to see a sidewinder in the wild, I followed the tracks to where they stopped at a creosote bush. I gingerly lifted the lowest branches and peered underneath. Nothing moved, nothing gave itself away. With a stick, I stirred up the sand in case the snake had curled up just under the surface. This too failed to get a response. Although the sidewinder has a reputation as a shy and mild-mannered snake, I felt that I had pushed my luck about as far as I dared, and I refrained from thrusting my hand into any of the numerous rodent holes where the snake might have hidden.

The last time I painted the interior of my house, I used shades of peachy pink, palest in the bedroom to ward off dark and heavy dreams, deepest in my studio to saturate my mind with color. Dabbing patches of paint here and there about the house, I sampled the menu of peaches and pinks, trying and discarding the evocatively named Rose Dawn, which turned bubble gum pink by incandescent light, and Renoir Red, a deep and erotic hue on the color chip but, at a larger scale, too reminiscent of barns. At last I settled on three shades that pleased my eye, and now when I walk from room to room, I pass from (god help us) Precious Peach through Apricot Lily to Peach Mimosa.

I quickly learned that few colors are as changeable as pink: it starts as peach in the early hours, pales to insignificance at midday, turns lavender as shadows lengthen. Dune fields are among the few natural landscapes that can match this changeableness. Most dunes are derived from quartz, and quartz is most commonly white, but sand most commonly is not, not paper white, anyway, nor as white as heavy cream. Yet few sands are as brown as caramel, either. Grab a handful, examine it, try to put a name to the color. Pretend you are a manufacturer of interior paints. There is no saying what color sand is unless you compare a sample with a Munsell color chart in the neutral environment of your laboratory. Outdoors, where no environment is neutral, the color of dunes varies throughout the day and maybe, for all I know, throughout the year. If I could sit in one place for a year, breathing in, breathing out, paying attention to nothing at all and everything at once, neither wishing for something to happen nor longing for surcease, I could find out.

Whether I could communicate what I learned is another matter. Last spring, I sat on an upper ridge of the Eureka Dunes for an hour or two, plopping myself down in warm sand as the sun approached the horizon and not leaving until the air was as cool as moonlight. The lower the sun sank, the more it gilded the milky sand. As the sun disappeared, shadows crept from indigo hollows, trailing lavender behind them. One raven flew over the dunes, then another. The only sound was the sibilant slash of their wings. Quite soon after that, the light was gone, as if nighttime were a cloak that the ravens had pulled over the earth. Even in darkness, the dunes refused to blacken like the surrounding hills but retained a glimmer of light—faint starlight, perhaps, magnified by grains of quartz and bounced back to the sky. I think you could travel a long way over dunes at night.

Mesquite Flat sand dunes with the Grapevine and Funeral Mountains in the background, seen in sunset's afterglow. Death Valley National Park, California.

Sunrise sky over the Granite Mountains, with the crest of Kelso Dunes in the foreground. Mojave National Preserve, California.

Undulating unnamed dunes in the eastern Mojave Desert with receding light marking the western crests of the dunes. California.

Undulating unnamed dunes in the eastern Mojave Desert glowing at dawn. California.

Flowering Borrego locoweed (Astragalus lentiginosus *var.* borreganus) *at sunrise with Providence Mountains in background. Kelso Dunes, Mojave National Preserve, California.*

Flowering Borrego locoweed (Astragalus lentiginosus *var.* borreganus) *at sunrise with Providence Mountains in background. Kelso Dunes, Mojave National Preserve, California.*

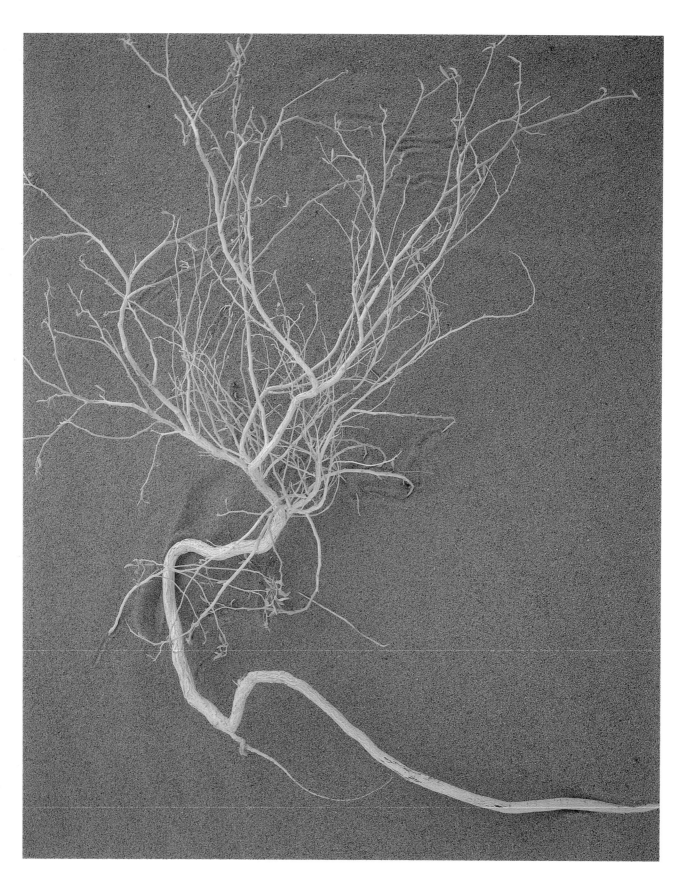

Croton (Croton californicus), *sun-bleached but alive. Mojave National Preserve, Kelso Dunes, California.*

Birdcage evening primrose (Oenothera deltoides) *at sunrise, with patterns in unnamed sand dunes. East Mojave Desert, Bureau of Land Management, California.*

Mesquite Flat sand dunes at sunrise, light on desert saltbush (Atriplex polycarpa) *and creosote bush* (Larrea tridentata). *Death Valley National Park, California.*

Mesquite Flat sand dunes with eroded, cracked mud patterns amid windblown sand patterns, at sunset. Death Valley National Park, California.

IV

I first learned of Racetrack Playa in a book, *In the Deserts of this Earth*. The author, a German naturalist named Uwe George, did not refer to the Racetrack by that name, nor would he divulge its location as other than "a remote mountain valley" somewhere in the Mojave Desert. I have an unfortunate tendency to forget what I read—glancing through George's book twenty years later, for instance, I have no recollection of him tying red ribbons around the legs of spadefoot toads—but I always remembered his description of a playa in the Mojave Desert where rocks, some weighing more than three hundred pounds, moved across the perfectly flat surface on their own, leaving long tracks etched into the silt. There was a rather unsatisfactory black and white photograph as confirmation.

I also retained an impression that the exact location of this place was a closely guarded secret. A ranger whom Uwe George pressed for information finally admitted, "Yes, yes, they're still there." According to George, the ranger yielded directions only with extreme reluctance. "I had the impression," George wrote, "that the minute he spoke he was sorry he had." Not only that, the ranger warned him on no account to stay overnight in the valley. Although I never expected to get there myself (how could I when its location was a secret?), I tucked this remote playa with its moving rocks in the back of my mind.

The second time I encountered the Racetrack was some years later at a university library. I happened to be leafing through a back issue of the *Geological Society of America Bulletin* and saw to my astonishment a black and white photograph of rocks and rock tracks. The photograph accompanied a scientific article, "Sliding stones, Racetrack Playa, California," written by two geologists named Robert Sharp and Dwight Carey. They made no mystery about the place at all. Although Racetrack Playa was well off the ordinary tourist routes, it was clearly not taboo or prohibited. I tucked this information in the back of my mind, too.

The Racetrack at sunrise: two boulders show their trails across the dry lakebed. Death Valley National Park, California.

The Racetrack at sunset: zigzag patterns on the dry lakebed, with Ubehebe Peak in the background. Death Valley National Park, California.

As far as Sharp and Carey were concerned, the only mystery about the Racetrack was how the rocks moved. They worked at the problem off and on for a number of years. Visiting the playa at intervals, they monitored the changing positions of thirty labeled rocks. (They gave each one a girl's name, as well, but this was not essential to the science of the project.) The smallest rock was not quite three inches in diameter, the largest about fourteen inches across. No two behaved exactly alike. In one winter, ten stones sidled away from the steel stakes that marked their position, while the rest sat unwavering like quail on their nests. In some winters, no stones moved, or only one or two did. The rock that rambled farthest, an eight-ounce cobble named Nancy, moved 659 feet in one season. Over the seven years of the study, Nancy traveled 860 feet. This is not a long distance—a desert tortoise could do as much in a day—but it is a remarkable achievement for a rock.

Neither Sharp nor Carey were lucky enough to see the stones move, but by careful study of the tracks, they learned a surprising amount about the process, such as the fact that some stones slide and others roll. Smooth stones, the geologists learned, are like boats without keels and cannot steer a straight course. Their tracks are smooth, whereas those left by rough stones are straight. Although rolling stones gather no moss, sliding stones push mud ahead of them, forming a levee to either side. Some get going so fast (perhaps as much two or three miles an hour) that, as Sharp and Carey put it, they "toboggan up on the outside edge of a curve." The geologists could recognize these speedsters by the extra mud they shoved to the outside edge.

Still more years passed before I finally got to the Racetrack myself. It was a cloudy, windy day as I set out from Beatty, Nevada, and I wished fervently for hurricane-force winds and heavy rain. Not until I got to the playa, of course—no point in being washed away in a flash flood. But if I wanted to see rocks skating across the flats, I would have to put up with a little discomfort because the combination of circumstances that sets them in motion apparently involves rain and wind: enough rain to slicken the clay surface of the playa and enough wind to overcome the initial resistance of the rocks. No one has ever seen the rocks move, or if they have, they've kept it to themselves. I'm not certain why I thought I deserved to be an exception. In any event, I was not. The air was quite still by the time I arrived, and the clouds had dispersed.

The biggest rocks and most impressive tracks are a fair distance from the road but I did not have to walk far to see the tracks of pebbles. As in Uwe George's photograph, the tracks were gray and blurry, like skid marks on pavement. Some went straight for long stretches, then swerved. Others curved like meandering streams, and some of these crossed over one another again and again, like ropes in diagrams of knot-tying. A twig tumbling in the wind had inscribed a pattern of parallel lines, one line per branchlet. After rainstorms, the surface of the playa had cracked into polygons as it dried. It looked as if someone had pressed an endless sheet of chicken wire onto the mud. I thought I could discern at least two overlapping systems of polygons, presumably from different storms.

I walked and walked and walked. Distances seem farther on flat surfaces and in clear air. When I finally got to the biggest rocks, I was in no way prepared for what I saw. That wind could slide a walnut-sized pebble across a slick, wet surface had not exhausted or even tried my imagination. But it was beyond me to conceive of a boulder the size of a hassock scraping a curvilinear track several hundred feet long. The hassock rock was the very picture of solid immobility; without tools—a sturdy wedge, a hammer, and some kind of lever—I doubted that I could budge it an inch. I recalled Uwe George and his encounter with the ranger: "Yes, yes, they're still there." As if they might have escaped. A three-hundred-pound rock might make a lot of progress on a seasonal or annual basis, but until it gets wings, it will in all likelihood still be there, wherever "there" might be.

The largest rocks, scattered tens or hundreds of feet apart, remained catatonic as long as I kept my eye on them, but their manic phase was plainly inscribed on the clay. The tracks variously converged and diverged and struck off for the great unknown and curved back on themselves and swerved and bent at crazy angles. Like ice skaters practicing individual routines, no two rocks left identical or even similar trails. I suppose that heavier rocks require stronger winds to initiate movement than lighter ones, and that rocks of different weights slide at different speeds in the same wind. During a given storm, therefore, not all rocks would move, and those that did would travel in the same direction but not necessarily the same distance. Repeated storms, driving different rocks in various directions, would account for the diversity of tracks. Maybe that is what appealed to me most about Racetrack Playa: that each rock had its own story to tell about where it had come from and what it had been doing. They were no longer just rocks. Possession of histories made them individuals to me.

I found one interesting exception to the rule of nonconformity. At its southeast edge, the playa abutted a limestone ridge that was the source of the rocks. Here a group of nine rocks had evidently fallen onto the playa in close proximity to one another, then traveled side by side for tens of feet through the same storm, then stopped for an unknown length of time. In the next storm, they all traveled at right angles to their previous course for another couple of feet, then stopped again. In a third storm, they continued in their original direction and slid another ten or twenty feet before halting.

Like all limestones, these nine rocks originated in an offshore environment as innumerable marine organisms died and drifted to the ocean bottom. Their own weight and that of the water above them eventually compressed their remains into solid rock. After the ocean retreated, tectonic forces lifted the limestone into a small, tiptilted mountain range, and erosive forces broke it into pieces. Gravity, working in fits and starts, nudged the pieces onto the playa. Anywhere else in the world, the fallen rocks would have stayed put. On Racetrack Playa, however, it's an ill wind that blows no rock some good, and by the time I saw them, the nine rocks had already started to write their own individual stories. Quite possibly they would travel and write until they wore themselves away.

V

For a number of years, it seemed that everyone I knew had seen a kit fox in the wild except for me. My friends Will and Kathy encountered one at the Eureka Dunes. They had retired for the night when they heard an animal banging among their pots and pans. When they beamed a flashlight in the direction of the noise, a kit fox looked up briefly, then continued its tour of their equipment. Another friend had a similar encounter in northern Mexico after grilling fish over a campfire. A choking sound awakened her in the middle of the night. The moon was full, and poking her head out of the tent, she saw a kit fox gagging and coughing. She deduced that it had a fish bone stuck in its throat, and she tossed it pieces of bread in the hope that a wad of something soft and thick would dislodge the bone. The kit fox snatched up the bread and trotted away, leaving the story without a proper denouement. My favorite kit fox story appears in Edmund Jaeger's *Desert Wildlife.* During a moonlit walk on Searles Dry Lake, he reports, a kit fox "tried to play hide-and-seek with us, running before, alongside, and behind us for more than an hour like an affectionate little dog."

Like thousands of other people, I had seen a kit fox in captivity at the Arizona-Sonora Desert Museum in Tucson. At night, when the museum is closed and no visitors are around, the kit fox presumably emerges above ground to prowl around its naturalistic enclosure. By day, it sleeps in an underground den, one side of which is a pane of glass. When you push a button on the wall, a dim light reveals the fox asleep, its back pressed against the glass, its nose tucked into its paws. This is sweet, but a kit fox behind glass must be an entirely different animal than one met on its own terms in the wild.

When would it be my turn? I wondered. Kit foxes evidently played with all and sundry but never gave me so much as a nod. Why, even my supervisor, a hydrologist and geomorphologist who generally regards wildlife as beside the point, if not an actual nuisance, had seen a kit fox. Only once, but that was one more time than I had. On a field trip to Death Valley, he had camped "way the hell up Cottonwood Canyon." He awoke in the middle of the night, opened his eyes, and looked right at a kit fox that was staring into his face. Over the next several days, he noticed abundant kit fox sign—scats, prints, and dens. "Lots of kit foxes there," he told me.

From that moment, I was on my way to Cottonwood Canyon, mentally if not physically. Life being what it is, several months elapsed before I got there, and although much of interest happened once I arrived, I did not see or hear a single kit fox in Cottonwood Canyon, and if there were any kit-fox dens, I missed them.

As we left Cottonwood Canyon, my husband said, "You never did see your kit fox."

To tell the truth, I had forgotten all about it by that point.

Maybe that was what was required.

Two months later, he and I happened to spend a day and a night at a small Mojave Desert dune field. We arrived late in the afternoon and set up camp, then he went for a quick look at wildflowers before dusk. I clambered a short distance up the dunes before realizing that I was too tired to do much of anything but sit and watch the sun go down. So I did.

A painted lady, a burnt-orange butterfly edged with black and spotted with white, careened crazily around me, then settled on the sand about a foot away. The painted lady is one of the most common butterflies in the United States. At that very moment, other painted ladies doubtless careened around my garden at home. The butterfly beside me was likely a hill-topping male. Hill-topping is a guy thing: males go to high points to seek out mates and spar with rivals. Females, ever on the lookout for the most virile and vigorous jeans—excuse me, genes—traipse after them.

An involuntary movement of my arm startled the butterfly into motion again. On the sand where the butterfly had rested, not a grain was overturned. When we were children, my sister and I gave one another "butterfly kisses" by flicking our eyelashes against the other's open palm. That slight flutter was much stronger than the weightless tread of a butterfly. After a good five minutes, the butterfly forgave or forgot me, and came back to the same spot. He alighted, then turned precisely ninety degrees and opened his wings to catch the last rays of daylight. We were among the fortunate of the earth, he and I—common as grains of quartz, no doubt about it, but look where we ended our day.

As I returned to our camp at the edge of the dune field, my husband met me in the twilight and said, "I have an interesting wildlife observation to report."

"A smilodont?" The sighting of saber-toothed tigers is one of our running jokes.

"Kit foxes. Two of them."

He handed me the binoculars, then led me to an area of sandy hummocks and swales. We stopped about thirty feet away from their den. It looked like a typical fox den: on the surface were mounds of sand beside several good-sized holes; underground there was presumably a system of burrows and chambers. I saw only one fox at first, a cream-colored female with huge ears, a pointed snout, and dark, pleading eyes like the eyes of a baby harp seal. All that showed was her head; the rest was hidden behind a mound. Edmund Jaeger says that what the kit fox lacks in speed, "it makes up by an extraordinary ability to make quick turns and take advantage of any scant cover of brush or rocks." This fox was so well concealed that I would not have seen her if my husband had not pointed her out.

She stared at me for some time; then, as she grew accustomed to my presence, she relaxed her vigilance, closed her eyes, and seemed to drowse; like a cat, however, she filtered sounds as she slept, and the slight disturbance I made by kneeling on the sand got her full attention for a moment. After about ten minutes or so, another fox trotted up behind her and nuzzled her head, then stopped abruptly and sniffed the air. The foxes exchanged looks; perhaps it was my imagination that she gestured toward me with a slight movement of her head or eyes. In any case, the other fox, a male, darted around her to get a better look at me, which gave me a good look at him: slender body, tan fur, sticklike legs, and bushy, black-tipped tail. Unlike the gentle beauty of his mate's face, his had an anxious, pinched expression.

When it was nearly dark, I stood up rather stiffly and brushed the sand from my knees. The foxes attended closely to my movements. Both got to their feet, and one of them barked softly: *wuff*, a quick exhalation of air. The male trotted back and forth beside the den, even came a few feet toward me until he was satisfied that I was leaving, not attacking.

Next morning, I was awake at first light. Back at the den, the female was curled up like a cat on a mound of sand. Her head rested on her front paws, and her tail curved around to cover her nose. She opened her eyes, looked at me, then closed her eyes, apparently content to sleep as long as I kept a safe distance. A cynic, I thought, might very well say that if I wanted to watch a kit fox sleep, I could have done so more conveniently at the Arizona-Sonora Desert Museum.

Eventually I went about the business of being a naturalist—that is to say, no particular business at all except wandering and watching. Painted ladies took nectar from desert marigold, an abundant wildflower with yellow heads and gray leaves and stems. Black-throated sparrows and house finches called from distant shrubs. Sphinx-moth caterpillars fed on leaves of evening primrose with reckless disregard of the relative quantities of caterpillar and leaf. Lizard and kangaroo rat tracks inscribed the lower dunes with reckless disregard of legibility.

It was a miraculous morning. I stumbled across things I had read about but never expected to see for myself. On the sandy flat at the base of the dune field I discovered the tunnel of a burrowing wolf spider. Like tarantulas, to which they are related, burrowing wolf spiders hide in vertical tunnels and seize whatever insects tread too close. The opening of the tunnel was about a half inch in diameter; spider silk lined the interior and was especially thick just below the rim where reinforcement was crucial. Having done some digging of dune sand myself, I was impressed by the spider's achievement. (Of course, I had not had access to a supply of spider silk with which to batten down the grains of sand as I worked.) This spider, a female, was not only an engineer, she was also an artist. Around the rim of the tunnel she had created a decorative collar made of marble-sized balls and a few dried leaves and petals. The collar probably served some practical purpose such as concealing the burrow or keeping out windblown debris, but, like the chrysalis of a butterfly or the spines on a pollen grain, it was beautiful in itself as well. I tentatively touched one of the sand-colored balls; fashioned of dust and gossamer, it crumbled instantly to nothing.

Next, as I walked along a little-used dirt road, the hum of many bees attracted my attention to the roadbed, which was, I saw now, pocked with dozens of tiny holes, as if someone had driven a carpet tack into the ground over and over again. I dropped to my knees for a better look. An astonishing amount of activity was crammed into an area about three feet square. A few bees were engaged in digging holes, their bodies bent double as they scrabbled at the dirt. Many more bees zigzagged over the roadbed: they appeared to rapidly survey the holes, then select one, drop into it headfirst, and, after some preliminary wiggling, disappear underground. After a minute or two, sometimes longer, the bee backed out of the hole and flew away. Their business, I surmised, was to place pollen in a hole and lay an egg, or several eggs, on top of it. The eggs would hatch into larvae, the larvae would eat the pollen and pupate, and eventually the pupae would mature and emerge as adult bees. A few bees were building inch-high turrets around certain holes. The turrets, apparently constructed of nothing but dirt and spit, were strong enough to withstand a light tap from thumb and forefinger. Like the collar around the spider burrow, the turrets no doubt served some practical purpose, but what it might be I could not guess.

The longer I watched, the more I saw: undifferentiated frenzy sorted itself into discrete activities. Some bees, it appeared, preferred to steal a nest hole rather than go to the bother of digging one. Would-be thieves risked nothing but a fight, and pairs of bees, presumably a thief and a rightful occupant, occasionally grappled and tumbled across the ground. Mixed among the bees were a quantity of flies that were the same size as the bees and somewhat resembled them. Although not as diversely occupied as the bees, the flies had their own carefully choreographed routine. To me it looked like this: buzz over the ground, locate an open bee nest, hover beside it, flick sand grains at the entrance with feet. One-two-three kick, just as in dance class. Why a fly would do such a thing was not at all obvious unless the occupant of the hole was expected to rush out and become a meal. Unfortunately for this hypothesis, bees and flies mutually ignored one another.

It remained a mystery to me until, back at home, I found an answer in an insect field guide. The flies were bee parasites that kicked their eggs into bee nests (with the possible exception of those nests protected by turrets). The fly eggs would hatch into maggots that would feed upon the living bee larvae. Ugh.

Ugh, but interesting all the same.

For sheer animal appeal, however, nothing matched the kit foxes, and after an hour or so, I wandered back toward the den. My husband was already there; in large-gestured pantomime, he motioned me forward with one hand, while putting a finger to his lips with the other. "Pups," he whispered, "five of them."

As Christina Rossetti wrote, "The birthday of my life is come."

There were indeed five not-yet-half-grown kits, their fur creamy in color and with the fuzzy quality of a soft undercoat. Their tails were black-tipped—as is the tail of every fox, whether red, gray, swift, or kit—but not as full as the adult's. The kits seemed as curious about us as we were about them, and as my husband walked around the area, bending over occasionally to examine a plant, they scrambled to the top of a mound and stood in full view to watch him. "What a fascinating guy," they seemed to say. I had always thought so myself; now I had confirmation.

Something startled the kits, and two of them popped into the den. Gradually, their eyes and ears emerged above the rim. How well they already understood the kit-fox hiding game. Checking on the young 'uns, I suppose, the male fox came out of the burrow. He watched us for a while, then curled up where he could keep an eye on us. Meanwhile, one of the kits began to play with a dead mouse. He mouthed it, shook it, tossed it over his shoulder, pounced on it, dropped it in front of a sibling, grabbed it and trotted away. Three siblings trotted after him. A fourth, greatly alarmed as a slight breeze ruffled the pages of my notebook, made a flying leap into the den. The fifth kit, as if she wanted attention or affection, nuzzled her father's nose and mouth, but the male fox repeatedly turned his head to one side or the other, gently refusing to oblige.

Entranced, we watched for an hour, maybe more. Two kits staged a mock battle, pawing one another, interlocking jaws, tumbling across the sand. The dead mouse had been abandoned by its previous owner, who lost interest as soon as his siblings stopped chasing him. Another kit, a different one this time, snatched up the dead mouse, and tossed it into a desert marigold. The kit pounced at the mouse from all sides, then stopped, stared, and waited as if expecting the mouse to run away. Two siblings ran up to join the game but possession seems to be nine-tenths of kit-fox law, and they were not allowed to play. One of them stalked off and picked up a small brown rock between his teeth. He carried it back toward the others, flopped down on his stomach and gnawed at the rock as if it too were a highly desirable toy. Every so often he glanced at the possessor of the mouse. His performance—for that is how it struck me—seemed to be an elaborate charade aimed at getting the mouse for himself, a kind of vulpine bait-and switch. Sibling number five, still wanting attention, pawed at her father's head, rolled down her father's snout, and wiggled on her back in front of her father's face, all to no avail. Father's attention was fully occupied with us, I'm afraid, and when we finally left, it was mainly to relieve him of the burden our presence imposed.

What do you do with the rest of the day when you have achieved your heart's desire before eight o'clock in the morning? I went back to camp and had breakfast. And then I set out across the dunes to look for another kit fox den.

VI

I have long maintained that the Rolling Stones have a song or phrase suitable to any situation. When we failed to find a kit fox in Cottonwood Canyon, I found myself thinking that if I couldn't always get what I wanted, at least I sometimes got what I needed. That was at the end of the day. At the beginning of the day, it looked as if I would get neither.

To maximize my chances of finding a kit fox, I wanted to spend as much time in Cottonwood Canyon as possible, which meant getting an early start. This proved unexpectedly difficult. First, my husband and I had to wait for the ranger station at Stovepipe Wells to open, then we waited again while another tourist harangued the duty ranger for ten minutes about back-country camping in national parks. The ranger, not knowing that we were professional botanists, occupied another ten minutes by kindly sharing a lot of information about wildflowers in Death Valley. Having finally gotten our back-country permit, we attached it to the dashboard of our truck and headed toward the Cottonwood Mountains over a dirt road. It was not a bad road—passenger cars could follow it for several miles—but it was not a good road, either, and my husband took it at a snail's pace.

My eagerness to arrive made the drive seem interminable until a zebra-tailed lizard leaped onto the roadway and dashed ahead of us as fast as it could run. My husband clocked it at exactly fifteen miles an hour. He turned to me and smiled. "Look at that little guy go," he said. I relaxed. After all, had not the Stones assured me that time was on my side?

"No es camino de altura velocidad," my husband added, quoting a sign once posted at frequent intervals along the major highway in Baja California: "not a high-speed road," as if you might not have noticed that potholes the size of volcanic craters constituted the greater part of the roadbed. On the road to Cottonwood Canyon, boulders and boulder holes enforced an unwritten speed limit. A field of boulders spilled from the mouth of Cottonwood Canyon, and the road had been arduously scraped around, over, and under the rocks. The boulder field, actually an alluvial fan, had been built up layer by layer during debris flows. A debris flow is a flood with teeth, a slurry of water, mud, gravel, and rock. From the looks of it, innumerable debris flows had sluiced down Cottonwood Canyon and spread across the fan, leaving a jumble of rocks.

Once the road entered Cottonwood Canyon proper, it deteriorated from a track into a rut and from a rut into a boulder garden. Finally my husband stopped the truck, got out, and walked ahead about a quarter of a mile. When he came back, he said that we had gone about as far as we could. There was just enough room where we stopped to turn around. "Let's camp here," he said, as I knew he would. This is how we usually find our campsites.

It was a lucky choice. Only five minutes from our camp, we found a bobcat lair. From the outside, it appeared to be a thicket of sturdy willows and cottonwoods; once you pushed past the drooping branches and coiled grapevines, however, you entered a leaf-wrapped room that was furnished with a sturdy horizontal branch about ten feet above the ground, perfect for catnaps and hidden reconnaissance. If you were a bobcat, you did not need to force your way in; you simply strolled through natural breaks in the underbrush of dead and living branches. The bobcat had marked two of these breaks by scraping away leaf litter and defecating on bare soil. One scat was fresh enough that I involuntarily glanced up at the branches overhead. "Uh, please don't drop onto my neck."

A find like this fills my plate, and I like to retreat for a period of digestion—making notes, checking field guides, shaping memories—to keep from being overwhelmed. Generally, this is easy enough in the desert, which is a place where one thing happens, then, a great while later, another thing happens. Not in Cottonwood Canyon. An avalanche would have been more restful. Willows hummed with bees whose pollen baskets bulged with yellow pollen. Mourning cloak butterflies dangled from cottonwood catkins, absorbing nectar through their strawlike tongues. Monarch butterflies, surprisingly large and deceptively fast, practiced tangos among the trees. A female Costa's hummingbird climbed into the sky, closely pursued by a male; at the peak of their ascent, they feinted like crafty boxers, then plummeted toward the ground in tandem, leaving no sky between them. Twenty feet up a cliff, as a bundle of sticks on a rock ledge resolved itself into a raven's nest, the raven flew close with something in her bill—the limp body of a lizard, perhaps, brought to feed her young. She veered off when she saw us and landed on a boulder some distance away. Only foolhardy birds go to their nests when a predator is watching. Ravens are no fools. *Urk, urk*, she said, then flew off.

We spent two days exploring Cottonwood Canyon and could have spent a lifetime. In a side canyon, we found outcrops of marble, a dry waterfall, and more bobcat scat. In the main canyon, a small stream came and went, flowing above ground for a mile or two, then disappearing under the sand. Wherever the water surfaced, cottonwoods and willows grew alongside, and migrating birds called from the treetops. To my astonishment, the birds obligingly centered themselves in my binocular lenses and all but shouted out their names—ruby-crowned kinglet, ladderback woodpecker, spotted towhee, lesser goldfinch, American robin, Oregon junco. In places, the stream spread across the canyon bottom, turning it into a marsh where rushes and cattails, flattened by floodwater, gave us a springy platform on which to walk. When willow saplings crowded together so densely that we could hardly press among them, we took to the canyon slopes and followed a track made by coyotes and bighorn sheep. From above, the gallery forest looked like a tossed green salad in a long, narrow bowl.

I would happily have stopped at any point to rest and gawk—maybe make a few notes, definitely let my excitement settle—but my husband insisted on continuing upstream. He is in general a severely practical person, but a romantic lurks in his soul, and geography has the power to lure it out of hiding. Three springs were marked on our map of Cottonwood Canyon, and he wanted to find them all.

The farther we went upstream, the earlier the season, until finally the cottonwoods were leafless, and willow buds were tightly wrapped in furry bracts. A raven's nest, unoccupied, filled the fork of two bare branches in a cottonwood tree. Other nests, too, were exposed, making it seem more like winter than spring, except that lizards rustled under last season's dead leaves.

At last we reached the upper end of the gallery forest. Standing on a butte where the canyon opened into an interior basin surrounded by hills, we could look for miles upstream. Blackbrush dotted the slopes but there was not a tree in sight. No trees, no springs. Plenty of water in the channel, though, most likely snow melt from higher elevations. My husband, loath to turn back without having completed his mission, walked upstream for another mile before admitting the obvious: the Cottonwood Springs, all three of them, were drowned in runoff.

I waited for him under a leafless cottonwood tree. Its knobby, white twigs nearly touched the ground, a curious phenomenon because cottonwood trees are typically self-pruning. As the trunks elongate, the lower branches break off, littering the soil with twigs and broken limbs and reminding the savvy camper to sleep well away or risk being crushed in the night. The classic cottonwood, therefore, has a tall trunk and arching limbs borne well overhead. But there I sat, my head literally in the tree tops as if I were Alice in Wonderland, psychedelically elongated and able to peer into bird's nests.

Suddenly I got it. I was in the treetops because the ground was higher than it used to be, and the ground was higher than it used to be because in the not-too-distant past—maybe twenty or thirty years earlier—a debris flow had surged downstream, burying the tree to its neck, or at least to its waist, in rocks, sand, and dirt. And not only this tree: the debris flow had traveled the length of the canyon, swirling around some trees, uprooting or breaking others. A debris flow, like well-mixed mortar, is thin enough to flow and thick enough to hold its shape. Because the slurry of rocks and mud does hold its shape, what remains after the moisture evaporates is a terrace. Every old tree that had survived the Cottonwood Canyon debris flow was now half-buried in terrace like fence posts in cement. No wonder the birds were so easy to see: the debris flow had brought the tree tops within easy viewing distance.

Debris flows can be immensely destructive. Judging from the weathered logs strewn here and there in Cottonwood Canyon, a good ten percent of cottonwood and willow trees had been ripped out. Debris flows can be rejuvenating, too. In the decades since the Cottonwood Canyon debris flow, fourwing salt-bush, indigo bush, Mormon tea, and other shrubs had colonized the debris-flow terraces, setting in motion a chain of events that had not yet come to a halt. Once the shrubs became established, kangaroo rats and other rodents dug burrows among their roots. As the shrubs grew large enough to produce flowers and fruits, birds, ground squirrels, and rabbits began to frequent the terraces. The presence of small animals attracted predators such as kit fox, great horned owl, coyote, and bobcat.

These same processes—colonization, reproduction, predation—go on in the desert at large as well, but at a different temporal and spatial scale. A creosote bush can live for several thousand years, and all that time it occupies space and consumes water and soil nutrients. Every year, thousands of seeds are pro-duced in its vicinity, but as long as the creosote bush maintains its hold, few if any can get started on that particular spot. Same with animals, more or less; the undisturbed desert might seem empty, but in reality it is fully occupied. Think of Beverly Hills, California, where individual mansions are surrounded by acres of lawn: there appears to be plenty of open space, but that does not mean that you or I can inhabit it. I remember reading that when the Oklahoma Territory was opened to homesteaders on September 16, 1893, thousands of people raced across the border in wagons and on horseback to claim free land. A soldier blew a bugle to start the race. Debris flows function both as bugle and free land, simultaneously setting the eco-logical clock to zero and creating new space.

At the end of the day, we had a long trudge back to camp. I was hot and tired, and my feet seemed to believe that I had been dropping heavy weights on them. We passed two men sitting near the stream in the shade of cottonwoods. They were leaning back to back, and each had a pair of binoculars to his eyes. I felt a sudden twinge of envy or even grief. Why hadn't we done that, just relaxed and watched the world pass by?

After we left the gallery forest behind, a cottontail, the only one I had seen all day, dodged behind a log. Evidently the rabbit population had taken a nosedive during the last drought, and although bunnies reproduce like, well, bunnies, it would probably be several seasons before cottontail numbers returned to normal. As the sun sank, black-throated sparrows sought out high spots from which to sing their evening song, which was just the same as their morning song but sadder somehow. *Deet, deet, deedleedee,* one called in high, clear tones, and from fifty yards away another replied, *Deet, deet, deedledee.*

The next day my husband and I lurched out of the canyon and down the alluvial fan toward the highway. As we drove away, I looked back at the Cottonwood Mountains. That's when my husband reminded me that I had not seen a kit fox. It did not matter. Although I had not gotten what I wanted, my needs had been amply fulfilled, just as the Rolling Stones had promised.

VII

Deserts are by definition places where people are apt to run out of water, yet knowledge of this possibility does not keep it from happening. If you hunted hard enough, you could probably find a sad story behind every waterhole and spring in the Mojave Desert: tales of springs reached too late or not reached at all, of sulfury springs and briny springs, of springs where the water was fine but the food ran out and springs where there was absolutely nothing to do but wait to be rescued.

The Forty-Niners, crossing the desert on their way to the California gold fields, became inadvertent protagonists of many sorry tales about dwindling food supplies and unlocated springs. At what is now known as Bennett's Well, the Bennett–Arcan party waited for twenty-six days while William Manly and John Rogers, provisioned with a ten-day supply of jerky and sixty dollars in change, went in search of civilization and enough food to enable the others to make it out alive. By the time Manly and Rogers returned with beans, wheat, meat, and oranges, one of the would-be rescuees, probably crazed with boredom, had struck out on his own and was later found dead. His canteen lay on the ground beside him. It was empty, of course.

Down the road from Bennett's Well is Tule Spring. A lifeline for early travelers, these and several other springs rise along the east base of the Panamint Mountains wherever groundwater is forced to the surface by a layer of impermeable bedrock. At Tule Spring, arrowweed, an upright shrub with narrow, gray leaves, grows thickly around the water, hiding it from view. To see the spring proper, you must penetrate the thicket, which turns out to be a kind of maze complete with curving paths and blind alleys. Arrowweed makes such an effective windbreak that no air stirs inside the thicket; mosquitoes thus find it all the easier to locate and attack the occasional intruder. The path to the center of the maze ends at a stagnant pool about the size of a dinner plate. Converging overhead, the arrowweeds cast deep but unrefreshing shade

Badwater, a heavily salted pool, offers a morning reflection of Telescope Peak covered with snow. Death Valley National Park, California.

on the water, which looks more like crude oil. A suffocating cloud of mosquitoes hovers over the surface. Of course.

This is not a place where I would choose to spend twenty-six days or even twenty-six hours. Twenty-six minutes was about right—enough time to get to the center of the maze and back, enough time to deduce from scat and tracks that cottontails come to Tule Spring for shade and mesquite pods, and that bobcats come for cottontails.

After living in the desert more than half my life I have become a kind of connoisseur of water and all things wet. Tadpoles, frogs, and garter snakes, for instance, please me inordinately, and I find it hard to ignore any body of water, even puddles remaining after storms. In September 1997, an unusual tropical storm dumped a couple of inches of rain in a wide swath across the Mojave Desert. A month later, you could still find large puddles in low-lying, shaded places. Hoping to find tadpoles or aquatic insects, I examined one such puddle that had collected under a railway trestle. What I found was a plethora of coyote tracks in mud and a female mallard paddling rapidly away from me.

Even when my physical thirst for water is slaked, my psychical thirst remains, and I suppose it will as long as I live in the desert. My friend Ray evidently feels the same, especially about hot springs. On a Mojave Desert field trip almost twenty years ago, a group of us happened to drive by the public bath house in Tecopa Hot Springs. "Shall we?" Ray asked. It was a rhetorical question; he had already slowed and was turning into the parking lot. The chance that he would have passed without stopping was about as great as the odds of beating the house in Las Vegas.

No mixed bathing was allowed. The men disappeared into the men's bath house. I grabbed a towel and some clean clothes and headed for the women's. There were two rooms, one furnished with benches and clothes hooks, the other containing a deep, tiled pool. The water in the pool was distinctly hot—not scalding, just hot enough to sting for a moment, as when you flick your wet fingertip on a steam iron. Sitting on the edge, I eased myself into the water, letting it clasp my ankles, rise up my back and belly, cup my breasts, and encircle my shoulders. By the time my feet touched the gravelly bottom, my very bones felt warmed and soft. I let my head loll back onto the water. The roof was louvered, and through the open slats I could see sky and cloud in strips. Wind rattled the louvers and blew the clouds.

A woman about my own age joined me in the pool. She filled her cupped hands with water, then wet her hair until it clung in long, thick, brown hanks to her neck and shoulders. Sunlight and shadow passed over the pool as clouds flew across the sun.

"Isn't it wonderful," she said

An old woman shuffled from the dressing room to the pool area. Leaning on the railing, she lowered herself into the water. If she had been dressed in street clothes, I would have thought of her as fat; in the pool, however, her movements had the heaviness, dignity, and grace of one accustomed to her body and its needs. After a few minutes, another elderly woman, evidently a friend, joined her. They stood side by side and talked about recipes and husbands while I read the book I had brought, a paperback called *Turtle Diary*. One of them, having examined the photograph of a man and woman on the cover, asked, "Is that a love story, dear?" I shook my head and explained that the book was about two people who free a pair of sea turtles from the London Zoo.

The woman with the long brown hair got out of the water and went into the dressing room. When she came back, she was wearing a green leotard, and her eyes and lips were freshly made up. Standing in the sunlight, she toweled her hair, then shook it loose. Stroke by stroke she brushed it, her gestures archetypically feminine. From the dressing room a woman's voice asked, "Does anybody mind if I sing?" No one objected, and her voice was strong and clear and resonant

Eventually I remembered that my co-workers might be waiting for me outside. I toweled off, dressed, and ran a comb through my wet hair. So lazy I could hardly move, limbs still floating in warm water, I opened the door and stepped into the wind. The men were waiting patiently in the van, as damp and clean as scrubbed puppies. "Did you enjoy yourself?" Ray asked.

"Look at her," someone said. "She didn't just enjoy herself, she had a religious experience."

Near the hot springs, groundwater feeds a large marsh called Grimshaw Lake. When I was there one recent May, it was a place of bright green reeds and brilliant blue water and innumerable birds. A marsh wren chided from the midst of rushes; I never did see her, but I could tell when I was getting close to her nest by the increased volume and rate of her scolding. Redwing blackbirds sang from the tips and tops of reeds and road signs. Their light-hearted, joyous *conk-a-lee* might be commonplace to some people, but living in the desert I can never get enough of it. I heard a yellowheaded blackbird, too, calling creakily as if his larynx needed a shot of Three-in-One Oil. My field guide describes the song as "a choking, scraping noise made with much apparent effort." This particular bird, I might add, also imitated a gas-powered lawnmower running out of fuel.

Birds that were not singing or scolding went about the never-ending business of finding food. Medium-sized shorebirds called avocets walked back and forth in the shallows and jabbed the water with their long, curved bills. Egrets, elegant in white feathers and black legs, stood perfectly still at water's edge. A pair of eared grebes, a species of diving waterfowl, disappeared below the surface for long periods as they hunted for small fish. Although the grebes stayed well away from the shore, I could see (with the help of binoculars) the fan of golden feathers from which they get their name. On a distant green island, about fifteen white-faced ibis—a large, cranelike bird—probed the mud with long, curved bills. Their dark plumage had a purplish-black sheen. Something, a barking dog, perhaps, or a passing truck, disturbed them, and the entire flock took off, legs dangling and vertical wings beating the air. They looked like black angels ascending to heaven.

Grimshaw Lake lies in a basin surrounded by low, white hills. Numerous seeps and springs arise on the hill slopes and trickle downhill to the marsh—green swatches and green ribbons on white damask. The hills are actually the remnants of an ancient lake terrace. Lake Tecopa filled the entire basin during the Pleistocene. The fauna of the lake included a species of pupfish that occurred nowhere else. At the end of the last Ice Age, the lake gradually evaporated, stranding pupfish in warm springs on the terraces. The fish survived into modern times and would be there still except for one circumstance: when public and private bath houses were constructed at Tecopa Hot Springs, no provision was made for pupfish. The species is now extinct.

When I enjoyed that long soak in the public bath house, I didn't know about the pupfish. Now that I do know, I am certain that hot baths, no matter how relaxing or therapeutic, cannot compensate for the loss.

Other species of pupfish still exist in other Mojave Desert drainages, as at Ash Meadows in southern Nevada. Ash Meadows is not only a refuge for pupfish, it is an extensive marsh, a wonderful wetland and a waterful wonderland. It was named for the scattered ash trees and for the meadowy swaths of saltgrass and rush, all fed by groundwater. Water rises to the surface everywhere in Ash Meadows—in ditches, at roadsides, on low ground, and especially along a chain of arid hills, where dozens of springs bubble up into stunningly clear pools. Much of this water originated high in the mountains of southern and central Nevada as rainfall and snowmelt. Some evaporated, some was consumed by plants and animals, and some percolated into the soil and followed the dictates of gravity and slope until it reached base level in Ash Meadows.

This was not a rapid journey: the water that flows into the springs today is about 8,000 years old. During the time it took a molecule of water to trickle underground from the Spring Mountains near Las Vegas to Ash Meadows, the above ground world changed. Valley-bottom lakes dried up, leaving flat playas and sheets of sand. Pupfish and other little fish were marooned in widely scattered pools and streams. Treeline crept up mountains by a thousand feet or so, and piñon-juniper woodland retreated from the valley floors. The Shoshone Indians arrived and made Ash Meadows their winter home. Lieutenant George Wheeler, assigned to his first scientific and topographic survey of the American West, put Ash Meadows on a map, if not the map. Las Vegas was founded (with the dubious results we see today). Would-be miners passed through Ash Meadows on their way to the California gold fields. Some came back to homestead in Ash Meadows, having decided that there was more money in selling beef, vegetables, and fruit to miners than in mining itself.

The arrival of Anglos was almost the beginning of the end for Ash Meadows. When I first visited in

1983, cattle had grazed there for more than a century with the usual effects: grasses chewed to the nub, shit splattered everywhere, and earth trampled hard wherever cows could rest in the shade. Meadows had been plowed and turned into cultivated fields. Springs had been diverted from their natural channels and used to irrigate crops. After the fields had been abandoned, dust devils and wind storms had blown the topsoil away. Tamarisk, a tree native to Eurasia, had been planted for shade but had quickly escaped and invaded irrigation ditches, springs, and every other wet place. Crayfish, bullfrogs, and mosquitofish, all exotic species, were introduced to the pools and streams. They preyed upon the native pupfish, decimating the population. Groundwater pumping threatened to complete what exotic animals had started: the water table was sinking, and, as a result, some springs had dried up, eliminating several populations of pupfish, and other springs and populations were in danger.

In 1984, a large part of Ash Meadows—about 22,000 acres—became a National Wildlife Refuge. Groundwater pumping was made illegal, and the lengthy process of restoration was begun: removal of bullfrogs, crayfish, and other exotic animals, rediversion of streams from irrigation canals to their original channels, bulldozing of tamarisk stands. By the time of my most recent visit in 1998, restoration of Crystal Spring was virtually complete, and work had started on King's Spring.

Crystal Spring is a place where I could happily spend twenty-six days, with or without something to read. A well-constructed boardwalk takes the place of a trail, leading you from refuge headquarters to the spring and having the dual advantage of keeping your feet dry in mud season and preventing damage to the soft terrain. Vigorous mesquite trees grow near the boardwalk and far into the distance. Their seed pods are tight corkscrews that identify the trees as screwbean mesquite, a species that tolerates salty soils as long as they are wet enough. There is no question of the saltiness of this particular soil. Salt colors the ground white and crystallizes around the stems of dead grass. Coyote tracks too are rimmed with salt as if with hoarfrost.

The boardwalk ends at a circular pool about twenty feet across and perhaps four to five feet deep. The deepest part of the pool shades from turquoise to aquamarine, a wide beautiful eye that never closes.

Water enters from below ground with enough force that the surface of the pool shivers like jellied broth, even when there is no wind. In spring and summer, dragonflies whiz back and forth across the pool, and marsh wrens, cross at disturbance and anxious lest you find their nests, scold in sharp tones: *Tsuk-tsuk, tsuk-tsuk.* Pupfish nibble algae and devour mosquito larvae. These are Ash Meadows pupfish, and although hundreds can be seen here now, groundwater pumping had nearly eliminated them by the late 1970s.

Like all pupfish, the Ash Meadows pupfish are only a few inches long. Their deep bodies and frilly tails distinguish them from other small fish that also inhabit desert waterholes. Male pupfish are larger than females and flushed with blue. When sexually mature, the males stake out and guard individual territories in shallow water. I once watched a male pupfish simultaneously guard his territory from interlopers and court a female who was looking for love. He was one busy little fish. When I first saw him, he was hovering in the middle of his territory, which was about the size of a cereal bowl, and dashing at any male who came too close. But when a female pupfish entered his zone, his personality changed, and he became Mr. May-I-Show-You-My-Etchings. Bodies touching, they swam side by side as she toured his territory. He herded her (or followed her) through loops and figure eights for several minutes. As if they were a single fish, they wiggled together over sand bars and stones. Heaven knows what would have happened next if another male had not interfered. Mr. May-I-Show-You darted at the interloper, and as the two of them tussled and tumbled together, the female scooted away.

Before leaving Crystal Spring, I always stop to read an interpretive sign posted near the water. It quotes three schoolchildren. When asked why we should save species and natural places, they responded:

"Because we love them."

"Because we'll be lonely without them."

"Because we can."

The children were right on all three counts, especially the last. We do have the power to save species and natural places if we want. But do we have the wisdom? The Greek word *hubris* is sometimes translated as excessive pride or arrogance. For the ancient Greeks, hubris meant setting yourself above the gods: for example, ignoring their warnings in the mistaken belief that you controlled your own life. It is easy to play at being God, but, as anyone who keeps an aquarium or a fish pond can attest, it is far from easy to do it well. It was hubris that allowed people to build bath houses at the expense of the Tecopa Springs pupfish, hubris that encouraged ranchers to exploit Ash Meadows until it was nearly gone.

Hubris, like a chronic disease, goes into remission then flares up again. I encountered a bad outbreak on a warm May afternoon at the Amargosa River. Like many desert rivers, the Amargosa is mostly a wide, cobbled streambed that lacks only water. Near the Dumont Dunes, however, a short perennial reach supports a thriving population of pupfish. On this particular day, there was also a thriving population of movie folk from Fox Studios. Several dozen of them were assembled under a big blue canopy on a terrace beside the river. A canteen truck, a motor home, and two tractor-trailer rigs were parked nearby. Porta-potties had been set up some distance away. Every fifteen or twenty minutes, another Blazer or Bronco or Cherokee arrived at or departed from this base camp with two or three passengers.

Down by the river, a huge cylindrical tank had been erected on a hydraulic lift. There was enough room underneath for a truck to drive through. Painted on either side of the tank was a toll-free phone number and the words Rain for Rent. A wide-diameter hose led from the tank to the river, and a noisy generator supplied the power for pumping river water into the tank. Although there was a filter on the end of the hose, the holes in the filter were about a centimeter in diameter, easily large enough to admit a pupfish. A number of pupfish were swimming near the filter.

Two young security guards, both women, sat on lawn chairs in the shade of the tank. They told me that their job was to guard the generator and pump. The water in the tank was for sprinkling the roads to keep down the dust. They knew nothing about the pupfish. Presumably the proper permits had been applied for and granted.

No doubt permission had been secured from the proper authority, an administrator sitting at a desk hundreds of miles away. But that did not make it right. What next, I wondered. The Mojave Desert Café featuring pupfish bisque?

Living in the desert for a long time does odd things to your psyche. You start to believe that pupfish are more important than mineral baths, more important than movies, even. You then realize that yours is a minority point of view and that the history of most desert waterholes is actually a series of anecdotes illustrating how water can be wasted, misused, and abused. You develop an urge to conserve as much water as you can: sprinkling the lawn less often (or getting rid of it altogether), ignoring the dust on your car, turning off the tap when brushing your teeth. While you shower, you marvel that these molecules of water have been on their way to your fingertips and scalp for thousands of years, starting as raindrops plinking onto dirt and finally swirling through miles of pipe to spill from a spigot into your cupped hands. You recall the pupfish playing in Crystal Spring, regret those lost at Tecopa Hot Springs, and wish that as a species we had more wisdom, less hubris, when exercising the powers we command.

Sunrise on the sandstone walls reflected in the Virgin River. Virgin River Canyon, Beaver Dam Paiute Wilderness, Arizona.

Pickleweed (Allenrolfea occidentalis) *at dusk, with the Argus and Inyo Mountains in the background. Warm Sulphur Springs, Panamint Valley, Death Valley National Park, California.*

Salt formations in crystallized patterns with dawn's light on the Panamint Range reflected in the pools. Death Valley National Park, California.

Sunset over the salt formations of Bristol Lake, with a storm approaching over Lead Mountain. Bristol Lake, Mojave Desert, Bureau of Land Management, California.

Darwin Falls, above Panamint Valley, with lush garden of ferns and cattails. Death Valley National Park, California.

Opposite: Red sandstone canyon with stream runoff reflecting the sky. Red Rock Canyon National Conservation Area, Nevada.

Red sandstone formations rise from Mojave Desert with reflection in pothole pools, at sunset. Red Rock Canyon National Conservation Area, Bureau of Land Management, Nevada.

The remaining waterhole in Panamint Valley's dry lakebed, with fading light on Panamint Butte in the Cottonwood Mountains. Death Valley National Park, California.

VIII

My grandmother, who lived next door to us when I was a child, occasionally took the train from California to Oklahoma, and on virtually every trip, she spent hours sidelined in Barstow in the western Mojave Desert. Not that our small town was any great metropolis, or that we ourselves were people of much sophistication, but the impression I got was that you might as well be sidelined in hell.

The western Mojave Desert strikes many people this way, probably because the highways take you through some of the most monotonous country you will ever see, not excluding large portions of Texas. Unless, of course, you think that polka dots are interesting: the predominant shrub, creosote bush, grows across the plains at mathematically precise intervals, green spots equally spaced on a neutral background. The regular spacing comes about as plants send out their roots in search of the water that has trickled down into the soil after rains. This water exists as a kind of film that occupies the space among individual grains of sand or clay or silt. If creosote bush roots are to find enough water to sustain life, plants of more or less equal size require more or less equal volumes of soil, and through competition they space themselves at a more or less equal distance.

The result is monotony, and as you drive through the desert at sixty miles per hour, every mile looks just like the mile before and the mile to come. The highway seems to purposefully avoid any feature that

Joshua trees (Yucca brevifolia) *at sunset with summer "monsoon" storm. Mojave Desert, Grand Wash Cliffs, Mohave County, Arizona.*

might catch your attention: even the hills have been pushed so far back that you can hardly see them. To keep from falling asleep, you fiddle with the radio, bite your wrist, slap your face. You envy your companion, snoozing placidly in the passenger seat. You wonder if anything interesting will ever happen again. You ask yourself, Why was I born? Why must I suffer so?

This is the point where it is helpful to recall Edward Abbey's advice to park rangers: Get off your butts and range, or something to that effect. In other words, stop the car, open the door, step outside, look around.

Last spring, driving toward Barstow along an endless stretch of highway, a virtual Mobius strip of creosote bush and asphalt, sheer boredom led me to try Abbey's strong medicine. I decided to drive another three miles, pull off the road, cross to the opposite side, and wander around until I found five interesting things.

The search started on an unfortunate note because directly across the highway from where I parked was a junk pile. Did this count as a thing of interest? Certainly it counted as a diversion from monotony, and I made a list of the more readily recognizable contents: the rusted chassis of a car, two torn sofa cushions, a dish rack from a dishwasher, a macramé plant hanger, a dented truck fender, broken sheet glass, a cracked garden hose, the inevitable bedsprings, a fiberglass bathtub, pieces of tar paper, a telephone book, many unmatched shoes. In rural Africa, none of this would have been left lying around; every scrap would have found a use somehow somewhere. But this was America, and, more to the point, it was the western Mojave, the sort of desert that people scornfully regard as just desert (analogous, I suppose, to just a bug, just a book, or just a Jew), with the result that they feel free to scatter large and unsightly portions of their lives across the landscape.

Thinking these baleful thoughts, I continued my search. Four more things of interest. A wide swath of ground beside the highway had been cleared for a gas pipeline. Regular disking apparently kept the swath clean of vegetation except for one thing: the person who rode the tractor had spared an old Joshua tree, which now stood by itself in the middle of the otherwise bare right-of-way, a monument to someone's finer sensibilities. Two down, three to go.

As I walked around, I realized that the supposedly monotonous plant community was much richer in species than appeared to be the case at sixty miles an hour. In addition to creosote bush and Joshua tree, I noted winter fat, Mormon tea, big galleta grass, and Indian rice grass. These were perennials, the woody and herbaceous plants that can live for several to many years. Because it was spring, Mediterranean split grass, an annual weed that germinates after winter rains, was abundant, and because it was a good spring, there were also a number of annual wildflowers—fiddleneck, sand verbena, yellow evening prim-rose, brown-eyed evening primrose, locoweed, blazing star, and some teeny-flowered thing in the Mustard Family that I did not recognize. Sixteen species altogether, eight times as many as I had noticed while driving.

Item of interest number four: every Joshua tree served a number of purposes besides its own, which is of course to make more Joshua trees. When I lifted a branch of a fallen tree, tiny white termites that had been consuming the fibers fled into a pin-head-sized hole in the ground. Where pieces of bark had sloughed away from the branch, the wood was riddled with insect galleries, perhaps the work of carpenter bees or beetle grubs. Woodpeckers, which nest in cavities, had gouged fist-sized holes into the trunk before it fell. Good nest holes are always at a premium in a desert where trees are few, and after the woodpeckers abandoned the holes, other kinds of birds doubtless moved in. At the base of a living Joshua tree, a pack rat had built its midden, a pile of debris that hides and protects a network of burrows. The midden incorporated an example of just about every movable object in the vicinity: sticks, rocks, leaves, bark, glass, bottle caps, wire, and so on. Pack rats eat a variety of plant stuffs such as leaves, seeds, and cactus pads or joints. They also relish the succulent flowers of Joshua trees and other yuccas. This particular pack rat had evidently gone to some trouble to get at the flowers or fruits borne high above its nest, because it had trimmed off the sharp-pointed tips of the leaves, making a well-traveled path that spiraled up the trunk and along a major branch.

The fifth thing of interest was the relative proportions of dead, old, and young Joshua trees. Unfortunately, the ratios did not augur well for this particular population. Most of the living plants were tall and much-branched; I had no way to age them, but certainly they were old. Young plants were few; in fact, there were more carcasses than young plants. In the desert, a healthy plant population should have many young plants, some middle-aged plants, and a few old plants. That way, as the old plants die, there are enough young ones to replace them. At this site, however, the ratio of old to young was the reverse of ideal.

Woody plants in the desert play Lotto every year of their lives, and the oldest plants are the most experienced losers. One difficulty is that the desert is a place where you can count the raindrops but not count on them, which is to say that climate is extremely variable from year to year. After dry years, for instance, Joshua trees bloom poorly—fewer flower stalks per plant and fewer flowers per stalk than after a normal year (whatever that is). The yucca moth that pollinates the flowers (incredibly enough, she makes a ball of pollen at one flower, then carries it to another flower on a different plant and, using her chin, rubs the pollen onto the stigma, thus transferring male gametes to the female reproductive organ), this little moth has ups and downs, too, plentiful in some years, scarce in others; consequently the number of seeds produced also goes up and down.

But even given plenty of rain and ample pollination, Joshua trees might bear many clumps of tight-fisted white flowers that eventually produce fat, green pods containing lots of hard, black, triangular seeds, and all for naught. Perhaps the caterpillars of the yucca moth (she lays an egg or two in every flower that she pollinates as fair recompense for service rendered) eat many of the developing seeds, and pack rats get the rest. Perhaps some seeds survive but there is not enough rain to trigger germination. Perhaps a few seeds germinate but cows trample all the seedlings. Something of that nature had happened along that highway, and Joshua trees were dying out locally as a result.

So, although it was a lovely gesture to leave a lone Joshua tree in the pipeline right-of-way, it was also about as futile as stopping to save tarantulas intent on crossing a busy highway. Futile or not, we do such things anyway, and this is one of the things I like best about us as a species. We invest our hearts and minds in other creatures because, as little children have told us, we love them, because we would be lonely without them, and because we can.

Phacelia (Phacelia tanacetifolia) *flowering amid dried smoke tree* (Psorothamnus spinosus) *branches, with the Pinto Mountains in the background, at dawn. Joshua Tree National Park, California.*

Opposite: Phacelia (Phacelia tanacetifolia) *flowering amid dried burrobush* (Hymenoclea salsola) *stems. Coxcomb Mountains, Joshua Tree National Park, California.*

Joshua trees (Yucca brevifolia) *in bloom after a rainstorm. Joshua Forest Parkway. Arizona.*

Joshua Trees (Yucca brevifolia) *in winter fog with cinder cones and Cima Dome in background. East Mojave, Mojave National Preserve, California.*

Wilson Cliffs at first light, with Joshua trees (Yucca brevifolia) *in the surrounding Mojave desert. Red Rock Canyon National Conservation Area, Nevada.*

Opposite: Saguaro cactus (Carnegia gigantea), *unique to the Sonoran Desert, with Joshua tree* (Yucca brevifolia) *of the Mojave Desert. Arrastra Wilderness, Arizona.*

Joshua trees (Yucca brevifolia) *against a brilliant sunset. Joshua Forest Parkway, Arizona State Trust Land, Arizona.*

Joshua trees (Yucca brevifolia) *silhouetted against sunset. Cima Dome, Mojave National Preserve, California.*

Joshua trees at dawn after spring snow. Lee Flat, Death Valley National Park, California.

Flowering beargrass (Nolina bigelovii) *amid the granite boulders near the Santa Maria drainage. Joshua Forest Parkway, Mojave Desert, Bureau of Land Management, Arizona.*

IX

I am not the first to have noticed that something happens to rubbish as it ages. Ordinary and undistinguished objects become rare, or at least interesting. They acquire historical significance. We become grateful that people were so thoughtful as to scatter bits of their lives across the landscape. The Mojave Desert is rich in romantic litter of this sort. Even in the wildest places, a remote canyon, say, where you would swear no one had set foot before, you might find rusted cans so old that they crumble in your hands, or pieces of lavender-tinted glass that have become smaller and smaller with time but never disappeared altogether.

In just such a place, I found an old padlock of a type no longer in use. Its rusty brown body was about the size of a silver dollar, its hasp rectangular in cross section. Only a key with a cylindrical barrel could have opened the lock, but even had there been such a key, the hasp could no longer be shoved home, and the locking mechanism itself was hopelessly rusted. The day I figure out the appeal of such objects will be the day I stop collecting them. In the meantime, I remain a finder of things and a keeper of things found. I slipped the padlock into the pocket of my jeans, and I have it still, along with a jelly jar half filled with pieces of lavender-tinted glass, a handmade spike from an abandoned railway, two lids from old tins of cinnamon or some other spice, and some pieces of broken crockery, one of which, my favorite, is patterned in blue on white in a curvilinear, trellislike design.

Another time, having parked beside a dirt road in the eastern Mojave, I set out across the plain to look at some low hills that struck my fancy. A tongue of sand climbed the near side of the closest hill, and I wondered if it also spilled down the other side. Underfoot, the substrate was like cobblestone pavement except that no human labor had been involved in setting the cobbles side by side in a matrix of fine sand. As I walked, my feet pressed the pebbles deep into the sand, leaving them helter skelter instead of jigsaw-puzzle neat. A flock of sage sparrows played hide and seek with me. Twittering, they flew well ahead, then like sparse snowflakes they settled invisibly among the bushes. Rabbit tracks in long, straight lines angled across my path. I saw lots of animal tracks, in fact, but no human footprints, and why would there be? No reason to come here, which was no place in particular.

Rodents had dug burrows under the creosote bushes. Spider webs hung in the entrances of the numerous abandoned burrows. Fresh paw prints distinguished those few burrows in active use. Dish-shaped hollows in the sand held miscellaneous collections of seeds. Looking more closely, I saw that most of the "seeds" were actually empty husks. The hollows were evidently workstations where kangaroo rats or mice had accumulated and processed batches of seeds.

Certain rodents bury seeds in small caches, then sniff them out at a later date. This is hardly a fool-proof way to store food: other rodents can sniff out the seeds just as easily, and although the odds are against it, there might be enough rain to make the seeds germinate, which would deplete the larder in one fell swoop. This happens sometimes with creosote bush seeds. Creosote bush seeds are notoriously diffi-cult to germinate in the laboratory—it is not a matter of simply dropping them on filter paper and keeping them wet—and they are also quite picky out of doors. What they want is a substantial autumn rainstorm when the temperatures are neither too hot nor too cold. What they usually get is the chance to be eaten by some rodent or bird.

In 1998, however, creosote bush had won the lottery, probably when Hurricane Nora swept across the Mojave Desert, and creosote bush seedlings were scattered across the sandy plain, especially under big creosote bushes and also near small obstructions such as fallen twigs. Here and there, densely clustered seedlings marked former seed caches, now well beyond recovery. Most of the seedlings had a single pair of seed-leaves, shaped like tiny scimitars, and another pair of true leaves. None was more than an inch or two in height. Their roots, however, were already five or six inches long. Seedlings of desert shrubs tend not to waste energy on making foliage until they have secured a water supply. Even then survival is far from certain, and all but a small fraction of the seedlings here would probably die from drought stress when summer came. Some had already wilted and probably would not survive the winter.

As I neared the low hills, I stumbled over a broken wooden stake that was poking out of the ground. Odd, I thought, until I noticed more stakes on the hillslope. The stakes obviously marked the corners of a mining claim, and now I could see that what had appeared to be a jumble of rocks was actually a tailings pile outside the mine. Or what would have been a mine if it had not been abandoned: the shaft penetrated perhaps five feet into the slope and ended at a bedrock slab. I would have sworn that no person had ever visited this nondescript hill, but someone had not only beat me to the spot, but had worked a claim and camped here. Scattered between the creosote bushes were rusted food cans, among them the lid to a one-pound can of baking powder. A pound of baking powder—that is a lot of biscuits. There was a rusted and corroded skillet about a stone's throw away, as if the last batch of biscuits had been so bad that the cook had hurled the pan (along with a mighty oath) into the night.

Looking at the site as if I were a miner in the late nineteenth century, I saw that the sandy substrate would make a good sleeping pad, and that dead stems of white bursage could be piled together for a little campfire. The presence of kerosene cans suggested that lanterns had been available for working or reading after dark. There were no springs or creeks within miles and miles, so the miner doubtless hauled water, perhaps in barrels, meaning that there must have been a mule, or a wagon. I tried to imagine what it would be like to stay at this spot for months at a time, how the stars would change overhead at night, how the moon would wax and wane, how the labor of every day would be to chop and dig at that hillside, turning the continuous fabric of bedrock into a jumble of tailings.

I saw how lonely it must have been.

More to my taste are the small, abandoned settlements we call ghost towns. I could imagine living in Rhyolite, say, which in its heyday was the largest town in southern Nevada, a cultural center more impor-tant than Las Vegas just one hundred miles to the southwest. Population estimates of Rhyolite at its peak vary from 3,500 to 10,000. In any case, several thousand people lived in Rhyolite by the end of 1907. Eighteen months later, as one historian noted, "it was well on its way to total oblivion." Silver was the key to Rhyolite's prosperity and to its decline. The discovery of rich silver lodes in 1905 attracted miners, first of all, then families, doctors, bankers, school teachers, bakers, hoteliers, stock brokers, prostitutes, journalists —an entire cityful of people who needed clapboard houses, banks, a three-story schoolhouse, a large hotel, a jail, saloons, an elaborate railway depot. And then, when the panic of 1907 hit, Rhyolite's undercapitalized mines could not attract enough investments for further development. Public confidence in the mines lagged. Unemployed miners relocated to other districts. Businesses failed. The newspapers stopped publishing. By 1910, the population of Rhyolite had dropped by eighty percent, and in 1919 the post office finally closed.

Odd to imagine people buying bread at the bakery and making deposits at the bank, unaware that in another few years the streets would be empty, and that a decade after that, without human hands to patch and mend, the town would begin to tumble upon itself.

You can see it happen in successive photographs. Three walls of the bank fall, leaving only the splendid facade. Brick by brick the schoolhouse crumbles until only half of it remains. The jail, more sturdily constructed, loses only its roof and windows. Houses, reduced to piles of rubble and board, drop out of the pictures entirely.

I have visited the town site several times now, and each time a little more of Rhyolite lies on the ground than the time before—more walls collapsed, more beams fallen, more adobe blocks dissolved in rain and snow. Each time there is also less of Rhyolite than there was before, because people carry parts of it away. There must be several hundred pits where—what shall we call ourselves? finder-keepers? antique hunters? looters?—have dug for bottles, crockery, tins, and anything else of financial or sentimental value. Now what remains is mostly broken or badly rusted or both: bottle necks, tin cans, bedsprings, stovepipes, barrel staves, washtubs, bottle caps, and of course an infinite number of pieces of glass, glass being one thing that time can fracture but not obliterate.

Now that its human inhabitants have left, the desert is reclaiming Rhyolite for its own. Pack rats make ingenious use of the ruins, improvising shelter from broken concrete blocks, weathered boards, and assorted rocks. Shadscale, a kind of saltbush, has colonized rubbish piles all over town. Rabbits take cover among the shrubs, and in spring, sage sparrows sit in them and sing. Sometimes their song sounds like *Tck? Chirpy-chirp,* and sometimes like *Ch-chirpy-dee? Chirpy-chirp.* Shadscale grows on the once-bare railroad bed, too, and in the borrow pit beside it. The railroad bed serves as a corridor for coyotes and bobcats, which come to hunt the rabbits and rodents. Rock wrens bob and trill from rubble piles. Ravens scavenge in basements that have been open to sky for decades.

There is something reassuring about a ghost town; it is good to know that the marks humans make on the landscape do not last forever. Grasses and weeds grow between the floor boards; roads disappear into brush; banks, schools, and hotels disintegrate. Plants and animals thrive where people did not and maybe should not. Only our technology lets us enjoy the desert in such large numbers. We invented the padlock, but they, it turns out, have the key.

Selected Bibliography and Recommended Reading

Abbey, E. 1968. *Desert Solitaire: A Season in the Wilderness.* McGraw-Hill, New York.

Bowers, J. E. 1993. *Shrubs and Trees of the Southwest Deserts.* Southwest Parks and Monuments Association, Tucson, AZ.

Bowers, J. E. 1998. *Dune Country: A Naturalist's Look at the Plant Life of Southwestern Sand Dunes.* University of Arizona Press, Tucson.

Collier, M. 1990. *An Introduction to the Geology of Death Valley.* Death Valley Natural History Association, Death Valley, CA.

Cowles, R. B. 1977. *Desert Journal: Reflections of a Naturalist.* University of California Press, Berkeley.

Ferris, R. S. 1974. *Death Valley Wildflowers.* Death Valley Natural History Association, Death Valley, CA.

Goldsack, D. E., M. F. Leach, and C. Kilkenny. 1997. "Natural and artificial 'singing' sands." Nature 386:29.

Hanson, J. and R. B. Hanson. 1997. *Fifty Common Reptiles and Amphibians of the Southwest.* Southwest Parks and Monuments Association, Tucson, AZ.

Henry, J. D. 1996. *Living on the Edge: Foxes.* NorthWord, Minocqua, WI.

George, U. 1979. *In the Deserts of this Earth.* Harcourt Brace Jovanovich, New York.

Jaeger, E. C. 1961. *Desert Wildlife.* Stanford University Press, Stanford, CA.

Lingenfelter, R. E. 1986. *Death Valley and the Amargosa: A Land of Illusion.* University of California Press, Berkeley.

Munz, P. A. 1962. *California Desert Wildflowers.* University of California Press, Berkeley.

Nabhan, G. P. and J. Cole (eds.). 1988. *Arizona Highways Presents Desert Wildflowers.* Arizona Highways, Phoenix, AZ.

Patera, A. H. 1994. *Rhyolite: The Boom Years.* Western Places, Lake Grove, OR.

Sharp, R. P. and D. L. Carey. 1976. "Sliding stones, Racetrack Playa, California." Geological Society of America Bulletin 87:1704–1717.

Sharp, R. P. and A. F. Glazner. 1997. *Geology Underfoot in Death Valley and Owens Valley.* Mountain Press, Missoula, MT.

Soltz, D. L. and R. J. Naiman. 1978. *The Natural History of Native Fishes in the Death Valley System.* Natural History Museum of Los Angeles County, Los Angeles, CA.

Van Dyke, J. C. 1980. *The Desert.* Peregrine Smith, Salt Lake City, UT.

Zwinger, A. H. 1989. *The Mysterious Lands: A Naturalist Explores the Four Great Deserts of the Southwest.* E. P. Dutton, New York.

Acknowledgments

Projects like this cannot be accomplished without the assistance of many people.

I would like to thank my editor, Robert Morton, for his trust and guidance, and Bob McKee for his elegant design.

Park rangers Alan Van Valkenburg and Charlie Callagan at Death Valley National Park, and Chief Ranger Chuck Ward at the Red Rock National Conservation Area near Las Vegas, desert rats Eldon Hughs and Dennis G. Casebier of California shared their intimate knowledge of the landscape. Fellow photographers Larry and Donna Ulrich, and Jeff Foott shared camps, great dinners, and secret locations.

Special care for the processing of my images was accomplished under the watchful eyes of Gary Riedmiller, William Snyder, and Mary Findysz of Tucson's Photographic Works. Charles Allen of Fuji Film contributed to the success of this project by monitoring processed film for quality.

My being able to do a book project with Jan Bowers has been a special pleasure. She has been the silent collaborator on two previous books, supplying scientific names and enthusiasm while making my work seem credible. It is rare, indeed, to work with someone who is both a gifted writer and a superb naturalist.

I want especially to thank Robert Schaefer of Robert Schaefer Inc. for introducing me to Ron Leven and Dwight Lindsey of Schneider Optics. Their generosity in supplying the very latest in Schneider's optical achievements allowed images never before possible to be included in this book.

Thanks to all.

Technical note

My lenses for this book were the latest lens designs from Schneider Optics, mounted on my Arca Swiss monorail field camera. To further lighten my load, I used a Fuji Quickload holder with 4 x 5" Fuji Velvia film.
Here is a list of my lenses:

   58mm Schneider Super Angulon XL
   75mm Schneider Super Angulon
   110mm Schneider Super Symmar Aspherical
   120mm Schneider Makro Symmar HM
   120 mm Schneider Super Symmar HM
   180mm Schneider APO Symmar
   270mm Schneider G Claron
   400mm Schneider APO Tele-Xenar

For camera support I used three different models of Gitzo tripods, including a light weight carbon-fiber model for backpacking. For conditions requiring more stability, (in windy weather or with long lenses), I used a Kirk leg support attached to my heaviest Gitzo to provide a solid platform.   —JD